International Symposium on (e,2e), Double Photo-ionization and Related Topics and 18th International Symposium on Polarization and Correlation in Electronic and Atomic Collisions

Donostia-San Sebastian, Spain
30 July – 1 August 2015

ISBN: 978-1-7138-2955-3

Printed from e-media with permission by:

Curran Associates, Inc.
57 Morehouse Lane
Red Hook, NY 12571

Some format issues inherent in the e-media version may also appear in this print version.

Donostia International Physics Center (DIPC)
Paseo Manuel de Lardizabal, 4
20018 Donostia-San Sebastian (Gipuzkoa)
Spain

Phone: +34 943 01 53 68
Fax: +34 943 01 56 00

http://dips.ehu.es

Additional copies of this publication are available from:

Curran Associates, Inc.
57 Morehouse Lane
Red Hook, NY 12571 USA
Phone: 845-758-0400
Fax: 845-758-2633
Email: curran@proceedings.com
Web: www.proceedings.com

International Symposium on (e,2e), Double Photo-ionization and Related Topics

18th International Symposium on Polarization and Correlation in Electronic and Atomic Collisions

PROGRAM

Donostia – San Sebastián
July 30th – August 1st 2015

INTERNATIONAL SYMPOSIUM ON
(e,2e), DOUBLE PHOTO-IONIZATION AND RELATED TOPICS
and
18th INTERNATIONAL SYMPOSIUM ON POLARIZATION AND CORRELATION IN ELECTRONIC AND ATOMIC COLLISIONS

The *International Symposium on (e,2e), Double Photo-ionization and Related Topics* and the *18th International Symposium on Polarization and Correlation in Electronic and Atomic Collisions* are organized together in Donostia – San Sebastián (Basque Country, Spain), from July 30th to August 1st 2015. Both symposia are official satellite meetings of the XXIX International Conference on Photonic, Electronic, and Atomic Collisions (ICPEAC 2015), held in Toledo (Spain) on 22 –28 July 2015. The two symposia are merged into a single event with a high-level scientific program, covering a broad range of hot topics in atomic and molecular physics.

Organizers

Donostia International Physics Center DIPC

Centro de Física de Materiales CFM
CSIC-UPV/EHU

Summer Courses University of the Basque Country UPV/EHU

Contact and further information

Web address: http://e2epol.dipc.org
Email: e2epol@dipc.org

BOOK INDEX

COMMITTEES

Chairmen

- Ricardo Díez Muiño
  Centro de Física de Materiales CSIC-UPV/EHU and
  Donostia International Physics Center DIPC
  Donostia – San Sebastián (Spain)

- Nikolay M. Kabachnik
  Skobeltsyn Institute of Nuclear Physics,
  Lomonosov Moscow State University
  Moscow (Russia)

International Scientific Committee

- Ugo Ancarani (Université de Lorraine, Metz, France)
- Lorenzo Avaldi (CNR-IMIP, Rome, Italy)
- Klaus Bartschat (Drake University, Des Moines IA, USA)
- Jamal Berakdar (Martin-Luther-Universität Halle-Wittenberg, Germany)
- Nora Berrah (University of Connecticut, Storrs CT, USA)
- Igor Bray (Curtin University, Perth, Australia)
- Michael Brunger (Flinders University, Adelaide, Australia)
- Xiangjun Chen (University of Science and Technology, Hefei, Anhui, China)
- James Colgan (Los Alamos National Laboratory, Los Alamos NM, USA)
- Alexander Dorn (Max-Planck-Institut für Kernphysik, Heidelberg, Germany)
- Reinhard Dörner (Goethe-Universität, Frankfurt am Main, Germany)
- Danielle Dowek (Université Paris Sud, Orsay, France)
- Omar Ariel Fojón (Instituto de Física Rosario, Argentina)
- Timothy Gay (University of Nebraska – Lincoln NE, USA)
- Alexei N. Grum-Grzhimailo (Lomonosov Moscow State University, Moscow, Russia)
- Nikolay Kabachnik (Lomonosov Moscow State University, Moscow, Russia)
- Anatoli Kheifets (Australian National University, Canberra, Australia)
- Murtadha Khakoo (California State University, Fullerton CA, USA)
- George King (University of Manchester, UK)
- Tom Kirchner (York University, Toronto, Canada)
- Noriyuki Kouchi (Tokyo Institute of Technology, Tokyo, Japan
- Don H. Madison (Missouri S&T, Rolla MO, USA)
- Fernando Martín (Universidad Autónoma de Madrid, Spain)

- Michael Meyer (European XFEL, Hamburg, Germany)
- Andrew Murray (University of Manchester, UK)
- Roberto Rivarola (Instituto de Física Rosario, Argentina)
- Michael Schulz (Missouri S&T, Rolla MO, USA)
- Rajesh Srivastava (Indian Institute Of Technology Roorkee, Uttarakhand, India)
- Al Stauffer (York University, Toronto, Canada)
- Emma Sokell (University College Dublin, Ireland)
- Masahiko Takahashi (Tohoku University, Sendai, Miyagi, Japan)
- Kiyoshi Ueda (Tohoku University, Sendai, Miyagi, Japan)
- Joachim Ulrich (Max-Planck-Institut für Kernphysik, Heidelberg, Germany)
- Jim Williams (University of Western Australia, Perth, Australia)
- Akira Yagishita (Photon Factory, Tsukuba, Japan)

Local Organizing Committee

- Maite Alducin Ochoa (CFM CSIC-UPV/EHU & DIPC, Donostia-San Sebastián)
- María Blanco-Rey (UPV/EHU & DIPC, Donostia-San Sebastián)
- Iñaki Juaristi Oliden (CFM CSIC-UPV/EHU & DIPC, Donostia-San Sebastián)
- Mohammed Ahmed Nosir (CFM CSIC-UPV/EHU, Donostia-San Sebastián).

LOCATION AND VENUE

Donostia - San Sebastián is located in the northeast of the Spanish Basque Country: 110 km from Bilbao, 500 km from Madrid, and 800 km from Paris. It is a charming medium-size city set in a picturesque countryside, on the edge of the Biscay Gulf and surrounded by mountains. A city of deep-rooted international tourist tradition dating back to the 19th century, San Sebastián offers quite a few tourist attractions, a varied choice of hotels and one of the most creative gastronomies in the world.

Further information on the city and the region can be found at:
http://www.sansebastianturismo.com
http://tourism.euskadi.net/

The scientific meetings would be held in the unique environment of the Miramar Palace that overlooks San Sebastián's Bay. Its rooms are home to the University of the Basque Country Summer Courses.

Access to the rooms of the symposia

The symposia will take place in the eastern wing of the Palace. Hence, we recommend to enter the building though the eastern side entrance. Please check it below:

Lunch

Lunch is included in the symposia fee for the three days of the conference (July 30th, July 31st and August 1th). Lunch will be served at the Hotel Barceló Costa Vasca (Avenida Pío Baroja 15, San Sebastián), which is at walking distance from the Miramar Palace. Please check below a recommended path from the Palace to the Hotel:

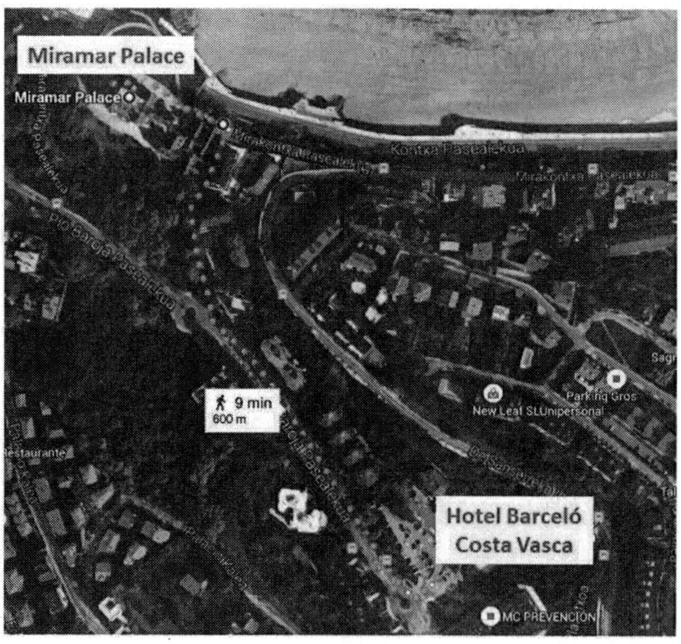

Conference dinner

The conference dinner will be held in Restaurante Ni Neu (Paseo de la Zurriola 1, San Sebastián, http://www.restaurantenineu.com), at the Kursaal complex. It will take place on July 31st at 8.30pm.

SCIENTIFIC PROGRAM

COMMON SESSIONS AND ACTIVITIES

REGISTRATION

Wednesday July 29th (17.30h-19.30h) and Thursday July 30th (8.00h-8.45h)

OPENING

Thursday July 30th:

8.45h-9.00h: Opening by Ricardo Díez Muiño and Nikolay Kabachnik

PLENARY TALKS

Thursday July 30th

Chairman: Ricardo Díez Muiño

29 **9.00h-9.45h: PL1.- Lorenz S. Cederbaum (University of Heidelberg, Germany)**
Interatomic (Intermolecular) Coulombic Decay and its Exploration by Free-Electron-Lasers

Friday July 31st

Chairman: Nikolay Kabachnik

30 **9.00h-9.45h: PL2.- Anatoli Kheifets (Australian National Univ., Canberra, Australia)**
Strong field atomic ionization with linear/circular polarized light: Spectra, Cusps and Time delay

Saturday August 1st

Chairman: Don Madison

31 **9.00h-9.45h: PL3.- Giovanni Stefani (Università Roma Tre, Rome, Italy)**
Direct and resonant double photoionization in solids

POSTER SESSION

Thursday July 30th

18.30h-20.00h: (Drinks and snacks will be served)

CONFERENCE DINNER

Friday July 31st

20.30h: Restaurante Ni Neu (Paseo de la Zurriola 1, Donostia – San Sebastián)

PARALLEL SESSION: ELECTRONS/IONS

Thursday July 30th

Chairman: Klaus Bartschat

35 **10.00h-10.30h: e-I1.- Daniel Fischer (Missouri S&T, Rolla MO, USA)**
Collision dynamics studied with a polarized MOT target
10.30h-11.00h: e-I2.- Leigh R. Hargreaves (California St. Univ., Fullerton CA, USA),
36 Unusual angular momentum transfer in electron-impact excitation of atoms and molecules

COFFEE BREAK

Chairman: James Colgan

37 **11.30h-12.00h: e-I3.- Klaus Bartschat (Drake University, Des Moines IA, USA)**
Non-perturbative calculations for electron-impact ionization of complex atoms
71 **12.00h-12.20h: e-O1.- Xueguang Ren (Phys.-Tech. Bundes., Braunschweig, Germany)**
Low-energy (67 eV) electron-impact induced interatomic coulombic decay of argon dimer
72 **12.20h-12.40h: e-O2.- Baoren Wei (Fudan University, Shanghai, China)**
Fragmentation mechanisms for Methane induced by 55 eV, 75 eV and 100 eV electron impact
73 **12.40h-13.00h: e-O3.- Stanislav Tashenov (University of Heidelberg, Germany)**
First observation of coherence in a highly charged ion

LUNCH

Chairman: Xiangjun Chen

38 **15.15h-15.45h: e-I4.- Béla Paripás (University of Miskolc, Hungary)**
Angle-dependent (e,2e) study of state-to-state interference between autoionizing states of He
39 **15.45h-16.15h: e-I5.- Nicholas Martin (University of Kentucky, Lexington KY, USA)**
Out-of-plane (e,2e) experiments on He autoionization levels
74 **16.15h-16.35h: e-O4.- Peter Koval (DIPC, San Sebastian, Spain)**
Iterative calculation of electron energy loss spectra
75 **16.35h-16.55h: e-O5.- Filippo Morini (Hasselt University, Diepenbeek, Belgium)**
The ground state nuclear dynamics of dimethyl ether in momentum space

COFFEE BREAK

Chairman: Béla Paripás

40 <u>**17.30h-18.00h:**</u> **e-I6.- James Colgan (Los Alamos Nat. Lab., Los Alamos NM, USA)**
Triple differential cross sections for the electron-impact ionization of laser-excited Mg: the perpendicular plane cross section

41 <u>**18.00h-18.30h:**</u> **e-I7.- Lorenzo Ugo Ancarani (Université de Lorraine, Metz, France)**
Sturmian approach for ionization processes: applications and perspectives

<u>Friday July 31st</u>

Chairman: Alexander Dorn

42 <u>**10.00h-10.30h:**</u> **e-I8.- Michael Schulz (Missouri S&T, Rolla MO, USA)**
Fully differential study of coherence and interference effects in ionization of H_2 by proton impact

43 <u>**10.30h-11.00h:**</u> **e-I9.- Shaofeng Zhang (Institute of Modern Physics, Lanzhou, China)**
"Double-slit" interferences observed in dielectronic transitions in collisions between hydrogen molecular ion and helium atom

COFFEE BREAK

Chairman: Masahiko Takahashi

44 <u>**11.30h-12.00h:**</u> **e-I10.- Masakazu Yamazaki (Tohoku University, Sendai, Japan)**
Towards making the molecular orbital movies by time-resolved (e,2e) electron momentum spectroscopy

76 <u>**12.00h-12.20h:**</u> **e-O6.- Mariusc Piwiński (Nicolaus Copernicus Univ., Toruń, Poland)**
Inelastic e-Cd and e-Zn collisions

77 <u>**12.20h-12.40h:**</u> **e-O7.- Chenzhong Dong (Northwest Normal Univ., Lanzhou, China)**
Influence of the higher order effects on the polarization and angular distribution of the radiation following electron-impact excitation process

LUNCH

Chairman: Allen Landers

45 <u>**15.15h-15.45h:**</u> **e-I11.- Allan Stauffer (York University, Toronto, Canada)**
Ionization of laser-excited atoms: the shape of the cross sections

46 <u>**15.45h-16.15h:**</u> **e-I12.- Andrey Surzhikov (Helmholtz-Institut, Jena, Germany)**
Interaction of twisted light with atomic and molecular targets

78 <u>**16.15h-16.35h:**</u> **e-O8.- Ladislau Nagy (Babeş-Bolyai University, Cluj, Romania)**
Projectile coherence -- the transition between the semiclassical and the quantum method

79 <u>**16.35h-16.55h:**</u> **e-O9.- Yew Kam Ho (Academia Sinica, Taipei, Taiwan)**
Electron-electron orbital entanglement in two-electron ions around the critical charge region

COFFEE BREAK

Chairman: Allan Stauffer

47 **17.30h-18.00h: e-I13.- Allen Landers (Auburn University AL, USA),**
Ion-momentum imaging of dissociative-electron-attachment dynamics in CO_2, N_2O, HCCH and CF_4

48 **18.00h-18.30h: e-I14.- Omar Ariel Fojón (Instituto de Física de Rosario, Argentina)**
Coherent emission from diatomic molecules: from femto- to atto-seconds

Saturday August 1st

Chairman: Michael Schulz

49 **10.00h-10.30h: e-I15.- Don Madison (Missouri S&T, Rolla MO, USA)**
Accuracy of theory for calculating 3-Body and 4-Body fully differential cross sections for electron-impact ionization of atoms and molecules

50 **10.30h-11.00h: e-I16.- Xiangjun Chen (Univ. of Science and Technology, Hefei, China)**
Three-body fragmentation of simple molecules induced by electron impact multiple ionizations

COFFEE BREAK

Chairman: Lorenzo Ugo Ancarani

51 **11.30h-12.00h: e-I17.- Alexander Dorn (MPI for Nucl. Physics, Heidelberg, Germany)**
A thorough study of Young-type interferences in (e,2e) on H_2 molecules with known spatial alignment

80 **12.00h-12.20h: e-O10.- Juana L. Gervasoni (Centro Atómico Bariloche, Argentina)**
Effects of the sudden electron-hole pair creation and of the life time of the residual hole on plasmon excitations in surfaces

81 **12.20h-12.40h: e-O11.- Károly Tõkési (Inst. Nuclear Research, Debrecen, Hungary)**
Optical constants of iron derived from reflection electron energy-loss spectra

LUNCH

PARALLEL SESSION: PHOTONS

Thursday July 30th

Chairman: Maria Novella Piancastelli

52 **10.00h-10.30h: ph-I1.- Emma Sokell (University College Dublin, Ireland)**
Coincidence photoelectron measurements following 2p photoionization in Mg

53 **10.30h-11.00h: ph-I2.- Kiyoshi Ueda (Tohoku University, Sendai, Japan)**
XFEL-induced ultrafast electron and molecular dynamics

COFFEE BREAK

Chairman: Emma Sokell

54 **11.30h-12.00h: ph-I3.- John Furst (University of Newcastle, Ourimbah, Australia)**
Alignment and Orientation of N_2^+

82 **12.00h-12.20h: ph-O1.- Miron Ya. Amusia (The Hebrew University, Jerusalem, Israel)**
One photon-two electron ionization in atoms and endohedrals – simplicity and complexity

83 **12.20h-12.40h: ph-O2.- Armin Scrinzi (Ludwig Maximilians Univ., Munich, Germany)**
Single- and double emission from multi-electron systems: How to compute IR double emission and solve the CO_2 mystery

84 **12.40h-13.00h: ph-O3.- Yuri V. Popov (Moscow State University, Moscow, Russia)**
New look at the strong field approximation in laser-matter interactions

LUNCH

Chairman: Kiyoshi Ueda

55 **15.15h-15.45h: ph-I4.- Alicia Palacios (Universidad Autónoma de Madrid, Spain)**
Decoding attosecond electron-nuclear dynamics in molecules by means of XUV-IR and XUV-XUV pump-probe schemes

56 **15.45h-16.15h: ph-I5.- Kirsten Schnorr (MPI für Kernphysik, Heidelberg, Germany)**
Electron Rearrangement Dynamics in Dissociating Iodine Molecules

85 **16.15h-16.35h: ph-O4.- Oksana Travnikova (LCPMR, Paris, France)**
Multi-step ultrafast fragmentation of the third row hydrides following K-shell excitation and ionisation

86 **16.35h-16.55h: ph-O5.- David Ayuso (Universidad Autónoma de Madrid, Spain)**
Ultrafast electron dynamics in phenylalanine initiated by attosecond pulses

COFFEE BREAK

Chairman: Alicia Palacios

57 **<u>17.30h-18.00h:</u> ph-I6.- Renaud Guillemin (Univ. Pierre et Marie Curie, Paris, France)**
Exploring deep-core photoionization: ion-electron correlation effects in the 1 to 10 keV x-ray region

58 **<u>18.00h-18.30h:</u> ph-I7.- Li Fang (University of Texas, Austin, TX USA)**
Photoionization induced fragmentation of glycine molecule and endohedral fullerenes $Ho_3N@C_{80}$ molecule

<u>Friday July 31st</u>

Chairman: Igor Bray

59 **<u>10.00h-10.30h:</u> ph-I8.- Artem Rudenko (Kansas State Univ., Manhattan, KS, USA)**
Electronic and nuclear dynamics triggered by ultra-intense soft and hard X-rays

60 **<u>10.30h-11.00h:</u> ph-I9.- Kyo Nakajima (JASRI, Hyogo, Japan)**
Photoelectron diffraction from laser-aligned molecules using an x-ray free-electron laser

COFFEE BREAK

Chairman: Tommaso Mazza

61 **<u>11.30h-12.00h:</u> ph-I10.- Olga Smirnova (Max Born Institute, Berlin, Germany)**
Attosecond Spectroscopy: from measuring ionization times to time-resolving chiral response

87 **<u>12.00h-12.20h:</u> ph-O6.- Alvaro Jiménez-Galán (Univ. Autónoma de Madrid, Spain)**
Time delay anisotropy in photoelectron emission from the isotropic ground state of helium

88 **<u>12.20h-12.40h:</u> ph-O7.- Ya-Wei Liu (Univ. of Science and Technology, Hefei, China)**
Optical oscillator strengths of the valence-shell excitations of molecular nitrogen measured by the dipole (γ,γ) method

89 **<u>12.40h-13.00h:</u> ph-O8.- Lin-Fan Zhu (Univ. of Science and Technology, Hefei, China)**
Squared form factors for the $A^1\Pi$ and $B^1\Sigma^+$ vibronic bands of carbon monoxide studied by high-resolution inelastic x-ray scattering

LUNCH

Chairman: Alexey Grum-Grzhimaylo

62 **<u>15.15h-15.45h:</u> ph-I11.- Tommaso Mazza (European XFEL, Hamburg, Germany)**
Circular dichroism in two-color multi-photon ionization of rare-gas atoms

63 **<u>15.45h-16.15h:</u> ph-I12.- Andrey K. Kazansky (DIPC, San Sebastián, Spain)**
Dichroism in two-color ionization of atoms by short pulses

90 **<u>16.15h-16.35h:</u> ph-O9.- Luca Argenti (Universidad Autónoma de Madrid, Spain)**
Modulation of Attosecond Beating in Resonant Two-Photon Ionization

91 **<u>16.35h-16.55h:</u> ph-O10.- Ralph Püttner(Freie Universität Berlin, Germany)**
The $1s^{-1}2s^{-1}$ and $1s^{-1}2p^{-1}$ double core-hole shake-up satellites in Argon

COFFEE BREAK

Chairman: Armin Scrinzi

64 **<u>17.30h-18.00h:</u> ph-I13.- Oleg S. Vasyutinskii (Ioffe Institute, St. Petersburg, Russia)**
Determination of spin-polarized H atoms produced in molecular photodissociation

65 **<u>18.00h-18.30h:</u> ph-I14.- Andreas Fischer (MPI für Kernphysik, Heidelberg, Germany),**
Molecular dynamics on laser-controlled transition states

<u>Saturday August 1st</u>

Chairman: Paola Bolognesi

66 **<u>10.00h-10.30h:</u> ph-I15.- Lorenzo Avaldi (CNR – Ist. Strutt. della Materia, Roma, Italy)**
Fragmentation of halopyrimidines and halouraciles by photoionization and ion impact

67 **<u>10.30h-11.00h:</u> ph-I16.- Laurent Nahon (Synchrotron SOLEIL, Saint Aubin, France)**
Gas phase targets in interaction with Circularly-Polarized Light: Molecular polarimetry and chiroptical effects in photoionization

COFFEE BREAK

Chairman: Lorenzo Avaldi

68 **<u>11.30h-12.00h:</u> ph-I17.- Christophe Nicolas (Synchr. SOLEIL, Saint Aubin, France)**
Momentum Exchange in Molecular Systems

92 **<u>12.00h-12.20h:</u> ph-O11.- Yaroslav Pavlyukh (Martin-Luther-Univ., Halle, Germany)**
Keldysh nonequilibrium Green's function vs. Feshbach projection operator approach for plasmon-assisted photoemission

93 **<u>12.20h-12.40h:</u> ph-O12.- Denis Iablonskyi (Tohoku University, Sendai, Japan)**
Interatomic Coulombic Decay Processes after Multiple Valence Excitations in Ne Clusters

LUNCH

POSTER SESSION

The poster session is scheduled on July 30th in the evening (6.30pm-8.00pm). Poster boards and scotch tape will be available since the morning of the first day. For each poster, the display space in the poster boards will be at least A0 size (841 x 1189 mm). The poster boards will be kept during the full duration of the symposia. We recommend the poster to be left at the boards until the end of the meeting.

LIST OF POSTER CONTRIBUTIONS

97 **P1.-** Photoelectron angular distributions and correlations in sequential two-photon double ionization by circularly polarized XUV radiation
Alexei N. Grum-Grzhimailo, Elena V. Gryzlova, Ekaterina I. Staroselskaya

98 **P2.-** Theory of ultrafast x-ray photoelectron diffraction
Shota Tsuru, Kyo Nakajima, Tokuei Sako, Takashi Fujikawa and Akira Yagishita

99 **P3.-** Two-photon triple ionization of lithium
James Colgan, M. S. Pindzola

100 **P4.-** Dressing Effects in the Attosecond Transient Absorption Spectra of Doubly-Excited States in Helium
L. Argenti, A. Jiménez-Galán, C. Marante, C. Ott, T. Pfeifer, F. Martín

101 **P5.-** Excitation of vibrational modes in the ionization of water molecule by XUV/X-ray radiation
Selma Engin, Jesús González-Vázquez, Inés Corral, Alicia Palacios, David Ayuso, Piero Decleva, and Fernando Martín

102 **P6.-** Photoionization time delays in molecular hydrogen.
R. Bello, Sebastian Heuser, A. Palacios, Matteo Lucchini, Lukas Gallmann, Claudio Cirelli, F. Martín, Ursula Keller

103 **P7.-** Mapping ultrafast dynamics of highly excited D_2^+ by ultrashort XUV pump - IR probe laser schemes
R. Bello, L. S. Martin, C. W. Hogle, A. Palacios, J. L. Sanz-Vicario, X. M. Tong, F. Martín, M. Murnane, H. C. Kapteyn and P. Ranitovic

104 **P8.-** Finite element DVR method for molecular single and double ionization by strong laser pulses
Denis Jelovina, Johannes Feist, Fernando Martín, and Alicia Palacios

105 **P9.-** Merging quantum chemistry packages with B-splines for the multichannel scattering problem.
Carlos Marante, Jesús González, Inés Corral, Markus Klinker, Luca Argenti, Fernando Martín

106 **P10.-** Ultrafast electron dynamics in phenylalanine initiated by attosecond pulses
F. Calegari, D. Ayuso, A. Trabattoni, L. Belshaw, S. De Camillis, S. Anumula, F. Frassetto, L. Poletto, A. Palacios, P. Decleva, J. Greenwood, F. Martín and M. Nisoli

107 **P11.-** Vibrationally resolved B 1s photoionization cross section of BF_3
D. Ayuso, M. Kimura, K. Kooser, M. Patanen, E. Plesiat, L. Argenti, S. Mondal, O. Travnikova, K. Sakai, A. Palacios, E. Kukk, P. Decleva, K. Ueda, F. Martín, C. Miron

108 **P12.-** Time delay anisotropy in photoelectron emission from the isotropic ground state of helium
S. Heuser, A. Jiménez-Galán, C. Cirelli, M. Sabbar, R. Boge, M. Lucchini, L. Gaallmann, I. Ivanov, A. Kheifets, J. M. Dahlström, E. Lindroth, L. Argenti, F. Martín, U. Keller

109 **P13.-** Phase Measurement of a Fano Resonance Using Tunable Attosecond Pulses
A. Jiménez-Galán, M. Kotur, D. Guénot, D. Kroon, E. W. Larsen, M. Louisy, S. Bengtsson, M. Miranda, J. Mauritsson, C. L. Arnold, S. E. Canton, M. Gisselbrecht, T. Carette, J. M. Dahlstrom, E. Lindroth, A. Maquet, L. Argentiz, F. Martín, A. L'Huillier

110 **P14.-** Temporal and spatial interferences in the laser-assisted photoionization of diatomic molecules
Diego I.R. Boll, Omar A. Fojón

111 **P15.-** Molecular-frame photoelectron angular distributions for carbon 1s photoemission of methyl iodide
H. Fukuzawa, S. Yamada, Y. Ito, T. Tachibana, T. Takanashi, Y. Sakakibara, K. Nagaya, T. Nishiyama, T. Sakai, M. Yao, M. Oura, N. Saito, M. Stener, P. Decleva, and K. Ueda

112 **P16.-** Subfemtosecond dynamics in dissociating core-excited CH_3I molecules studied with resonant Auger spectroscopy
T. Marchenko, G. Goldsztejn, L. Journel, R. Guillemin, O. Travnikova, A. F. Lago, D. Céolin, J.P. Rueff, R. Püttner, M. N. Piancastelli, and M. Simon

124 **P28.-** Correlation and quantum entanglement in Rydberg states of the helium atom
Yen-Chang Lin, Te-Kuei Fang, Yew Kam Ho

125 **P29.-** Effect of orthogonalization on total ionization cross sections by electron impact: Application to small molecules
S. Nehaoua, S. Houamer, C. Dal Cappello, M. Chinoune, A. Galstyan and A. C. Roy

126 **P30.-** Non-perturbative B-spline R-matrix with pseudo-states calculations for electron-impact excitation-ionization of helium to the $n = 3$ states
Oleg Zatsarinny and Klaus Bartschat

127 **P31.-** Mechanisms of Chiral Sensitivity in Electron-Molecule Interactions
J.M. Dreiling, F.W. Lewis, and T.J. Gay

128 **P32.-** Kinematically complete low-energy (e, 2e) study of neon: Internormalized triple-differential cross sections in 3D kinematics
XueGuang Ren, Sadek Amami, Oleg Zatsarinny, Thomas Pflüger, Marvin Weyland, Woon Yong Baek, Hans Rabus, Klaus Bartschat, Don Madison, Alexander Dorn

129 **P33.-** High resolution electron ejected spectra of He, Ne and Ar by high energy electrons
B. P. Marinković, J. J. Jureta, and A. R. Milosavljević

ABSTRACTS

PL	Plenary Lectures
e-I	Invited Talks (electrons/ions parallel session)
ph-I	Invited Talks (photons parallel session)
e-O	Oral Contributions (electrons/ions parallel session)
ph-O	Oral Contributions (photons parallel session)
P	Posters

PLENARY LECTURES

Interatomic (Intermolecular) Coulombic Decay and its Exploration by Free-Electron-Lasers

Lorenz S. Cederbaum

Department of Theoretical Chemistry, University of Heidelberg
D-69120 Heidelberg, Germany

How does a microscopic system like an atom or a small molecule get rid of the excess electronic energy it has acquired, for instance, by absorbing a photon? If this microscopic system is isolated, the issue has been much investigated and the answer to this question is more or less well known. But what happens if our system has neighbors as is usually the case in nature or in the laboratory? In a human society, if our stress is large, we would like to pass it over to our neighbors. Indeed, this is in brief what happens also to the sufficiently excited microscopic system. A new mechanism of energy transfer has been theoretically predicted and verified in several exciting experiments. This mechanism seems to prevail "everywhere" from the extreme quantum system of the He dimer to water and even to quantum dots. The transfer is ultrafast and typically dominates other relaxation pathways.

To exploit the high intensity of laser radiation, we propose to select frequencies at which single-photon absorption is of too low energy and two or more photons are needed to produce states of an atom that can undergo interatomic Coulombic decay (ICD) with its neighbors. ICD is an extremely efficient decay mechanism for excited systems which are embedded in environment. For the Ne2 dimer it is explicitly demonstrated that the proposed multiphoton absorption scheme is much more efficient than schemes used until now, which rely on single-photon absorption. Extensive calculations on Ne2 show how the low-energy ICD electrons and Ne+ pairs are produced for different laser intensities and pulse durations. At higher intensities the production of Ne+ pairs by successive ionization of the two atoms becomes competitive and the respective emitted electrons interfere with the ICD electrons. It is also shown that a measurement after a time delay can be used to determine the contribution of ICD even at high laser intensity. The study can provide a hint how the energy deposited by a FEL on one site in a medium can be transferred fast to the surrounding.

References

Work on ICD can be found in the ICD Bibliography:
http://www.pci.uni-heidelberg.de/tc/usr/icd/ICD.refbase.html

E-mail: Lorenz.Cederbaum@pci.uni-heidelberg.de

PL2

Strong field atomic ionization with linear/circular polarized light: Spectra, Cusps and Time delay

Anatoli Kheifets* [1] **and Igor Ivanov**† [2]

* Research School of Physics and Engineering, The Australian National University, Canberra, Australia †

†Center for Relativistic Laser Science, Institute for Basic Science, Gwangju, Republic of Korea

Synopsis We made significant progress in understanding the timing and structure information extracted from photoelectron spectra in strong field atomic ionization with short and intense electromagnetic pulses of varying ellipticity. The photoelectron spectra reveal peculiarity of strong field ionization in multi-photon, tunneling and over-the-barrier regimes and display a Coulomb singularity which can be removed by switching from linear to circular polarization. The time delay studies provide rigorous tests to various attosecond measuring schemes.

Photoelectron spectroscopy of strong field atomic ionization is a promissing tool to study the timing of the ionization process and to extract the structure information on the target. The time of the wavepacket release can be mapped on the photoelectron angular rotation [1] (circular polarization) or its kinetic energy [2] as well as the two-photon interference beating [3] (linear polarization). The structure information can be encoded in the transverse electron momentum distribution by filtering the momentum profile of the target orbital with a tunneling theory Gaussian [4].

The initial promising results of these techniques can now be scrutinized in more detail. It appears that the relative photoemission time delays between the $2s$ and $2p$ shells of Ne [2] and the $3s$ and $3p$ shells of Ar are in disagreement with accurate calculations with the full account of many-electron correlation and a non-perturbative treatment of strong field interaction [5, 6]. The attoclock measurements in He [1] cannot by unambiguously interpreted in terms of the tunneling time [7]. Predictions of the simple tunneling formula [4] are hampered by a Coulomb singularity [8].

In the latest set of strong field ionization measurements, a more thorough interpretation of the experimental data is made and a better agreement with theory is achieved. The cross-species time delay measurements by two-photon interference beating [9, 10] seem to be in a much better agreement with predictions of *ab initio* many-electron calculations [6]. Anisotropy effects in photoemission time delay can now be resolved experimentally and modelled from the first principles [11]. Relative time delay between single and double photoionization can also be accessed experimentally [12] and understood theoretically [13]. Attosecond spatial control of electron wave packet emission dynamics can be achieved by laser-sub-cycle two-dimensional electron-momentum mapping using orthogonal two-color fields [14].

The Coulomb singularity in transversed electron momentum distribution can be gradually removed by increasing ellipticity and switching from linear to circular polarization thus eliminating the cusp and restoring the tunneling theory predictions [15]. This effect allows to distinguish experimentally the tunneling and over the barrier regimes of strong field ionization [16]. The cusp structure is also modified by the account for non-dipole relativistic effects [17] which leads to a non-zero expectation value for the transverse electron momentum [18].

[1] E-mail: A.Kheifets@anu.edu.au

[2] E-mail: Igor.Ivanov@anu.edu.au

References

[1] P. Eckle *et al* 2008 *Science* **322** 1525

[2] M. Schultze *et al* 2010 *Science* **328** 1658

[3] K. Klünder *et al* 2011 *Phys. Rev. Lett.* **106** 143002

[4] N B Delone and V. P. Krainov 1998 *Physics-Uspekhi* **41** 469

[5] A. S. Kheifets and I. A. Ivanov 2010 *Phys. Rev. Lett.* **105** 233002

[6] A. S. Kheifets 2013 *Phys. Rev. A* **87** 063404

[7] L. Torlina *et al* 2015 *Nat. Phys.* **11** 503

[8] A. S. Kheifets and I. A. Ivanov 2014 *Phys. Rev. A* **90** 033404

[9] C. Palatchi *et al* 2014 *J. Phys. B* **47** 245003

[10] D. Guénot *et al* 2014 *J. Phys. B* **47** 245602

[11] S. Heuser *et al* 2015 *Nat. Phys.* submitted, arXiv:1503.08966

[12] E. P. Månsson *et al* 2014 *Nat. Phys.* **10** 207

[13] A. S. Kheifets, I. A. Ivanov and Igor Bray 2011 *J. Phys. B* **44** 101003

[14] Li Zhang *et al* 2014 *Phys. Rev. A* **90** 061401

[15] I. A. Ivanov 2014 *Phys. Rev. A* **90** 013418

[16] I. A. Ivanov *et al* 2015 *Phys. Rev. Lett.* submitted, arXiv:1503.04891

[17] I. A. Ivanov 2015 *Phys. Rev. A* **91** 043410

[18] C. T. L. Smeenk *et al* 2011 *Phys. Rev. Lett.* **106** 193002

PL3

Direct and resonant double photoionization in solids

Giovanni Stefani* [1]

*Dipartimento di Scienze, Università Roma Tre, Roma

Synopsis One photon in and two electrons out experiments on solids are reviewed in order to highlight their capability to put on sound experimental grounds some of the more widely used models for electron-electron correlation. Double Photoemission (DPE) and Auger Photoelectron Coincidence Experiments (APECS) are used to give experimental validation to concept such as the excitation-correlation hole, the hole-hole correlation energy and the one-two step ionization process .

Electron correlations are fundamental to account for many properties of matter in its different states of aggregation. The additional term in the Schrödinger equation describing them makes an analytical solution impossible and one must resort to approximations that restore simplicity through effective independent particle models. Hence the need for experimental validations of these models. Because of correlations, absorption of a single photon by a quantum system might be followed by emission of a pair of correlated electrons thus leaving behind a doubly ionized target. Such an event has been experimentally investigated only during the past few decades mostly following the availability of third generation synchrotron radiation sources and related multi-parameters experimental setups.

One photon in and two electrons out experiments on solids have demonstrated that correlation energy, exchange correlation-hole, single and double step ionization paths aren't purely conceptual models but rather experimental realities.

All of these experiments share a doubly ionized final state and two continuum electrons that are to be detected coincident in time and discriminate in energy and/or momentum. Two main reaction channels are possible: direct double photoionisation of the valence electrons (DPE), and core resonant double photoionization (APECS). Experiments connected to both channels will be reviewed in this talk.

DPE resolved in angle, i.e. discriminating in momentum the two final unbound electrons, have proven reality of the exchange correlation hole. It manifest itself in a depletion of the coincidences intensity whenever the electron pair has a negligible relative momentum [1].

APECS experiments consist in detecting in coincidence core photoelectrons and related Auger electrons and have shown an unprecedented discrimination power in the correlation energy U of the valence electrons. A case study on Cu will be presented and discussed [2].

The concept of dynamical screening of the core hole has been investigated by APECS from Ag and the crucial role played by the core-hole lifetime has been highlighted. By these experiments the attosecond time scale is accessible via the width of the energy sharing between electrons of the pair.

Angle resolved APECS (AR-APECS) experiments are also reviewed in the light of exploring dynamical changes in highly correlated systems, such as magnetic ones. In ferromagnets the parameters U, band width and magnetic exchange interactions are similar in magnitude and majority and minority spin electrons may exhibit dramatically different degrees of localization. We report on a case study by analyzing the Fe $M_{23}VV$ transition where we find that while decay channels that involve minority spin electrons are well described within an independent particle approximation, significant correlation effects must be included and accounted for the final states with two majority spin holes [3]. A spin dependent on-site electron correlation is then disclosed by these experiments.

Partial financial support through SIMDALEE2 *Sources, Interaction with Matter, Detection and Analysis of Low Energy Electrons 2* Marie Sklodowska Curie FP7-PEOPLE-2013-ITN Grant # 606988, is greatly acknowledged

References

[1] F.O. Schumann *et al* 2009 *Phys. Status Solidi B.* **246** 1483

[2] G. Di Filippo *et al* 2015 *J. Phys. Cond. Matt.* **27** 085003

[3] R. Gotter *et al* 2012 *Phys. Rev. Lett.* **109** 126401

[1] E-mail: stefani@fis.uniroma3.it

INVITED TALKS

Collision dynamics studied with a polarized MOT target

Daniel Fischer [1]

Missouri University of Science & Technology, Rolla, MO, USA
Max Planck Institute for Nuclear Physics, Heidelberg, Germany

Synopsis We report on the kinematically complete investigation of ion-impact induced single ionization of laser-cooled and state-prepared lithium atoms. The experimental combination of a magneto-optical trap with a momentum imaging spectrometer provides detailed insights in correlation and polarization effects in the collision dynamics.

Studying the dynamics of atomic collisions enhances our understanding of the fundamentally important few-body problem. In this respect, fully differential data of ionizing collisions represent an extremely sensitive test of theoretical models and calculations. For ion-impact ionization, such experimental data only become accessible since the development of reaction microscopes, often referred to as cold target recoil ion momentum spectroscopy (COLTRIMS). Here, the momentum vectors of the charged target fragments are measured providing the full kinematical information of ionizing scattering processes. In these experiments, the momentum resolution is limited by the initial temperature of the target atoms. Therefore, all fully-differential cross sections reported in literature have been obtained for helium targets, because this element can relatively easy be cooled to temperatures of 1K or even below using supersonic gas jets.

In this contribution, we report on the fully-differential data on ion-impact ionization for a target other than helium. The data has been obtained using a novel experimental development (MOTReMi), where a reaction microscope (ReMi) is combined with a magneto-optically trapped (MOT) target of lithium atoms [1]. Due to its electronic structure with two strongly correlated inner electrons and a single valence electron, lithium is an ideal candidate to examine electronic correlation and collision dynamics of simple few-body systems. In the MOT, lithium can be cooled down to temperatures below 1 mK resulting in a substantially improved resolution compared to conventional COLTRIMS experiments. Moreover, the target can be prepared in an excited state and even polarized using lasers.

The MOTReMi was integrated in the ion storage ring TSR of the MPIK in Heidelberg and single ionization of lithium in collisions with several projectile species ranging from 6 MeV protons to 1.5 MeV/amu O^{8+} has been investigated. Due to the high resolution and by means of optical excitation, for the first time initial state selective cross sections for ion impact ionization became available. Fully differential cross sections for the ionization of 1s, 2s and 2p electrons have been extracted which provide detailed information on interference and polarization effects in the scattering dynamics [2].

References

[1] D. Fischer, *et al.*, Phys. Rev. Lett. **109**, 113202 (2012)

[2] R. Hubele, *et al.*, Phys. Rev. Lett. **110**, 133201 (2013)

[1] E-mail: fischerda@mst.edu

e-I2

Unusual angular momentum transfer in electron-impact excitation of atoms and molecules

Leigh R. Hargreaves* [1]

*Department of Physics, California State University Fullerton, Fullerton, California, USA

Synopsis We present experimental and theoretical results for the circular polarization correlation parameters in the electron excitation of neon and hydrogen. The measurements compared well with results from the B-Spline R-Matrix model, but are in less good accord with other models.

In our recent study of polarization correlations in electron excitation of neon [1,2], we observed a negative angular momentum transfer to the excited state at small scattering angles. That is, a rotation of the excited state towards the scattering electron, rather than away from it. This observation was in contradiction to the long standing, but empirical, "propensity rule". Classically, the propensity rule can be justified as saying that, at small scattering angles, the scattering dynamics should be dominated by long-range repulsive Columbic interactions, which hence deflect the excited state charge cloud away from the scattered electron. However, it remained unclear whether the classical model was a sufficient description. The observation of the failure of the propensity rule thus demonstrated a clear breakdown of the classical model and demonstrated the quantum mechanical nature of the problem.

In addition, the measurements were compared with several theoretical calculations. It was demonstrated that the B-Spline R-Matrix approach (BSR) of Zatsarinny and Bartschat [3], when employed with a very large number of input states, was able to predict this behavior, whereas simpler distorted wave approaches continued to predict behavior consistent with the propensity rule.

Also apparent in the BSR calculations were predictions of strong oscillatory structures and sign changes in the angular momentum transfer at larger scattering angles, which again were not reflected in competing theoretical models. Such predictions are historically difficult to test experimentally, for the rare gases, as the experiments demand time-consuming coincidence detection of the scattered electron and vacuum ultraviolet fluorescence photon. Such experiments become challenging at large scattering angles as the scattering cross sections become small.

In this presentation, we discuss results from our ongoing electron-photon coincidence measurements of the circular polarization correlation parameters in neon, with a focus on testing the large angle predictions of the BSR. In addition, we give some preliminary data for polarization correlations in molecular hydrogen.

[1] L.R. Hargreaves *et al* 2012 *Phys. Rev. A.* **85** 050701
[2] L.R. Hargreaves *et al* 2013 *Phys. Rev. A.* **87** 022701
[3] O. Zatsarinny and K. Bartschat 2012 *Phys. Rev. A.* **86** 022717

[1] E-mail: lhargreaves@fullerton.edu

Non-perturbative calculations for electron-impact ionization of complex atoms

Klaus Bartschat[1]

Department of Physics and Astronomy, Drake University, Des Moines, Iowa 50311, USA

Synopsis We present an overview of recent calculations for electron-impact ionization, including ionization with simultaneous excitation, of several noble-gas atoms. The advantages, disadvantages, and limitations of various approaches applied to obtain these results are discussed.

In recent years, much progress has been made in the calculation of electron-impact ionization of relatively simple targets such as atomic hydrogen as well as helium, provided the latter remains in the electronic ground state of the residual ion, i.e., one of the 1s electrons can effectively be considered a spectator. For (e,2e) on atomic hydrogen, fully non-perturbative methods such as convergent close-coupling (CCC), exterior complex scaling (ECS), and time-dependent close-coupling (TDCC) have been applied with great success, and CCC, TDCC, as well as the B-spline R-matrix (BSR) with pseudo-states approach have also achieved impressive results for ionization without excitation of helium. Several of these methods were then employed, with similar success, to quasi-one-electron and quasi-two-electron targets (i.e., the alkali and alkaline-earth metals) by treating all inner electrons as an inert core represented by an effective potential.

Electron-impact ionization of more complex targets, such as Ne or Ar, or more correlated processes, such as ionization with simultaneous excitation of He, Mg, or Ca, is an even more challenging task, both experimentally because of low count rates and theoretically due to the fact that at least three electrons change their quantum state considerably. For problems like this, more approximate methods, based at least partially on perturbation theory, have been a common choice. Such approaches include the well-known distorted-wave ansatz, sometimes modified with a Gamov factor to approximately account for the post-collision interaction, methods such as 3CC or 3DW, in which the correct three-body asymptotic form of the wavefunction is employed while the quality of the short-range description might be sacrificed, and hybrid distorted-wave + close-coupling approaches. In the latter, the interaction between a fast projectile electron and the target is described by a first-order (DWB1) or second-order (DWB2) model. The initial state as well as the ejected-electron – residual-ion interaction, on the other hand, are described by an R-matrix with pseudo-states (RMPS) close-coupling approach.

Most recently, the non-perturbative BSR method was further developed and applied to both direct ionization [1] and ionization with excitation problems [2,3]. Here a projection approach is used to obtain the cross sections from excitation amplitudes of the positive-energy discrete (i.e., finite-range) pseudo-states through a mapping procedure of the pseudo-states to the physical continuum states of the ejected electron.

Figure 1 shows an example of a recent calculation for electron-impact ionization of Ne (2p) at the rather low incident energy of 65 eV [4]. While the agreement between the BSR predictions and experiment is by no means perfect, it is much better than for the DWB2-RM method. More examples will be discussed at the conference.

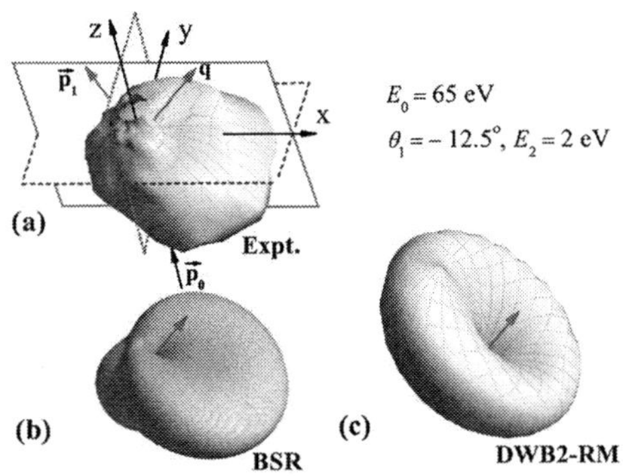

Figure 1. TDCS for ionization of Ne (2p) by incident electrons with energy $E_0 = 65$ eV [4], presented as 3D images. Panel (a) shows the experimental data, while panels (b) and (c) represent the predictions from the BSR and DWB2-RM models.

A significant part of this work was carried out in collaboration with Oleg Zatsarinny. Financial support from the United States National Science Foundation under grants PHY-1212450 and PHY-1430245, and the XSEDE allocation PHY-090031, is gratefully acknowledged.

References

[1] O. Zatsarinny and K. Bartschat 2012 *Phys. Rev. A* **85** 062709

[2] O. Zatsarinny and K. Bartschat 2011 *Phys. Rev. Lett.* **107** 023203

[3] O. Zatsarinny and K. Bartschat 2014 *J. Phys. B* **47** 061001

[4] X. Ren *et al.* 2015, *Phys. Rev. A* **91** 032707

[1] E-mail: klaus.bartschat@drake.edu

e-I4

Angle-dependent (e,2e) study of state-to-state interference between autoionizing states of He

B. Paripás[1] and B. Palásthy[2]

Institute of Physics, University of Miskolc, 3515 Miskolc-Egyetemváros, Hungary

Synopsis The electron impact excitation of the autoionizing states of helium, their subsequent decay into the same $He^{+}1s^{-1}$ final ionic state, and the interference of these processes have been studied. We have found some signs of the exchange interference of the $2s^2(^1S)$ and $2p^2(^1D)$ states in our previous measurements at large acceptance angles of spectrometers. Then we divided this accepted angular range into four smaller parts and measured the spectra sector by sector. Our evaluation method is based on the comparison of spectra measured at the critical primary energy and a few eV below or above it, where the scattered and ejected electron peaks are well separated.

Two autoionizing resonances with a common final ionic state can be made to interfere at a certain electron impact energy [1]. We concentrated on the possible state-to-state interferences of the $2s^2(^1S)$ and $2p^2(^1D)$ autoionizing resonances of He at energies 57.8 eV and 59.9 eV (decaying to the same simple $He^{+}1s^{-1}$ final state with E_F = 24.6 eV), and their possible interference. This state-to-state (exchange) interference (which is essentially differs from the Fano interference, which also can happen) occurs at a unique (critical) electron impact energy, where the energy of the scattered electron from one reaction path equals the energy of the ejected electron released along the other path and vice versa: in that case the scattered-emitted electron pairs are indistinguishable.

These electron pairs are observed in coincidence by two cylindrical mirror analysers. We have found some signs of the exchange interference of the above states in our previous measurements [2]. The weak reproducibility of these signs was partly attributed to the large acceptance angles of our spectrometers. So we divided the large angular range of one of the spectrometers into four smaller parts: forward sector with 44°-62,4° scattering angle, upward and downward 59,4°-120,6°, backward 117,6°-136°. Then we measured both the total and the coincidence spectra sector by sector.

Our evaluation method is based on the comparison of spectra measured at the critical primary energy and a few eV below or above it, where the scattered and ejected electron peaks are well separated (Fig.1.). At the critical energy the scattered and ejected electron peaks are shifted onto each other (e.g. the $2s^2(^1S)$ej and $2p^2(^1D)$sc peaks) and their possible interference results different peak shapes for both the total and the coincidence spectra.

To increase the accuracy, the spectra are evaluated by a computer code, using the Shore-Balashov parameterisation [3] of the peaks.

We have found that around the critical energy, a few tenths of eV modification in primary energy causes a significant change in the spectra, which can prove the state-to-state interference effect.

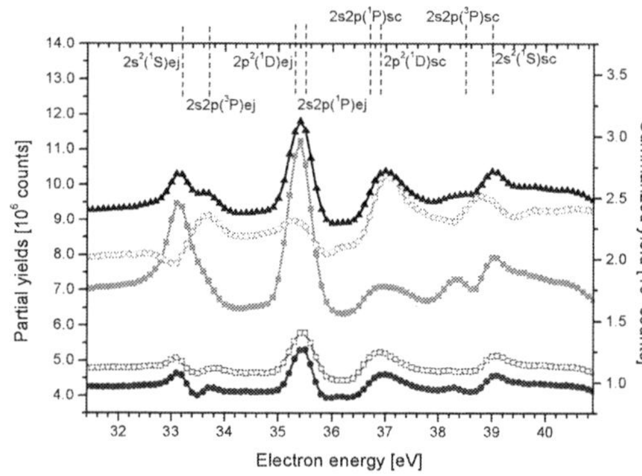

Figure 1. The measured spectra at 96.9 eV primary energy in the different angular ranges: forward (empty circle), upward (full circle), backward (full square), downward (empty square). The **sum of these spectra** (full triangle) is also shown. The position of the scattered (sc) and ejected (ej) electron peaks are shown by dashed lines.

This research was (partially) carried out in the framework of the Centre of Excellence of Mechatronics and Logistics at the University of Miskolc.

References

[1] J. P. V. den Brink, G. Nienhuis, J. van Eck and H. M. Heideman 1989 *J. Phys. B* **22** 3501.
[2] B. Paripás, B. Palásthy and G. Pszota 2014 *Eur. Phys. J. D*, **69**: 34
[3] B. W. Shore 1967 Rev. Mod. Phys. **39** 439.

[1]E-mail: fizpari@uni-miskolc.hu
[2]E-mail: fizpalb@uni-miskolc.hu

Out-of-plane $(e, 2e)$ experiments on He autoionizing levels

N.L.S. Martin* [1]

C.M. Weaver,* [2] **B. A. deHarak,**† [3] **K. Bartschat**‡ [4]

* Department of Physics and Astronomy, University of Kentucky, Lexington, KY 40506-0055, U.S.A.
† Physics Department, Illinois Wesleyan University, P.O. Box 2900, Bloomington, IL 61702-2900, U.S.A.
‡ Department of Physics and Astronomy, Drake University, Des Moines, IA 50311, U.S.A.

Synopsis
$(e, 2e)$ experiments on He autoionizing levels for 150 eV incident-electron energy are reported and compared with our previous experiments and calculations at 488 eV.

Up to a decade ago, all (e,2e) experiments were of the coplanar type where the ejected electrons were detected in the scattering plane formed by the incident- and scattered-electron trajectories. Theory was in quite good agreement with such experiments. About 10 years ago, so called out-of-plane experiments, where the ejected electron was observed out of the scattering plane, found results for direct (non-resonant) ionization that could not be explained by existing theories. In 2008 we carried out the first out-of-plane (e,2e) experiments and calculations on He $2\ell 2\ell'$ autoionizing levels for an incident energy of 488 eV [1]. It was found that second-order calculations were essential for a good description of the experimental results. We are currently carrying out similar experiments for an incident energy of 150 eV.

The geometry of our out-of-plane experiments is shown in Fig 1, for an incident electron momentum $\vec{k}_0$ and scattered electron momentum $\vec{k}_{sc}$. Plane I corresponds to a coplanar experiment, and plane II is perpendicular to the momentum transfer $\vec{K} = \vec{k}_0 - \vec{k}_{sc}$ and the scattering plane. Our experiments correspond to plane III, which is perpendicular to both the scattering plane and $\hat{k}_{sc}$, but also contains $\hat{K}$. This plane is selected by choosing the appropriate scattering angle. The energy loss for the He autoionizing states is ~60 eV and the corresponding momentum transfer for plane III is in all cases 2.1 au.

I will give a progress report on the 150 eV experiments and present our current results.

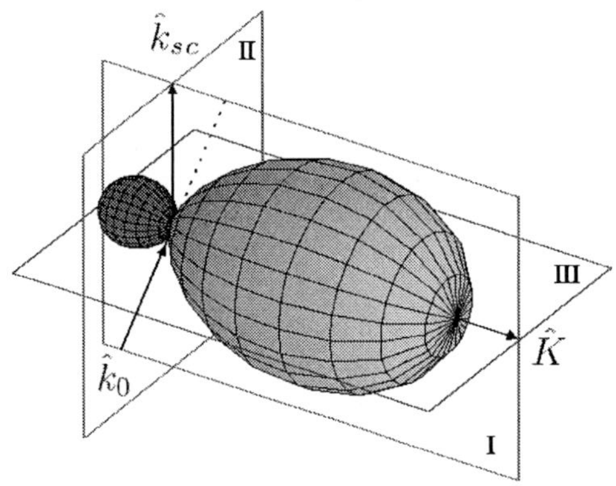

Figure 1. Geometry of the out-of-plane experiments.

This work was supported by the United States National Science Foundation under Grants No. PHY-0855040 (NLSM), PHY-1402899 (BAdH), and PHY-1212450 (KB).

References

[1] B. A. deHarak, K. Bartschat, and N. L. S. Martin, Phys. Rev. Lett. **100**, 063201 (2008).

[1] E-mail: nmartin@uky.edu
[2] E-mail: mike.weaver@uky.edu
[3] E-mail: bdeharak@iwu.edu
[4] E-mail: klaus.bartschat@drake.edu

e-I6

Triple differential cross sections for the electron-impact ionization of laser-excited Mg: the perpendicular plane cross section

James Colgan[1 1], G. S. J. Armstrong[1,2], M. S. Pindzola[3], A. J. Murray[4], K. L. Nixon[4], S. Amami[5], D. H. Madison[5], A. Stauffer[6]

[1] Theoretical Division, Los Alamos National Laboratory, Los Alamos, NM 87545, USA
[2] Department of Physics, Kansas State University, Manhattan, KS 66506, USA
[3] Department of Physics, Auburn University, Auburn AL 36849, USA
[4] School of Physics and Astronomy, University of Manchester, Manchester M13 9PL, UK
[5] Physics Department, Missouri University of Science and Technology, Rolla, MO 65409, USA
[6] Department of Physics and Astronomy, York University, Toronto, Ontario, Canada M3J 1P3

Synopsis We use the time-dependent close-coupling approach to examine electron-impact ionization of laser-excited Mg. We consider in particular the perpendicular plane geometry, for which a non-zero cross section was measured but a zero cross section had previously been theoretically predicted.

It has recently been shown that angular distributions for the electron-impact ionization of excited states of atoms may be measured by first exciting the target with a laser and then performing a 'usual' electron-scattering measurement [1, 2]. This was demonstrated for Mg and it was also quickly realized that various orientations of the excited state could be probed by changing the laser direction with respect to the incoming electron beam. The resulting angular distributions showed a clear dependence on the orientation of the Mg $3p$ state.

Recent theoretical work has reported analysis of the measurements using symmetry arguments [3] and reported calculations that compare with such measurements [4]. For many of the geometries considered, good agreement was found between time-independent distorted-wave calculations and measurement, but in the plane perpendicular to the incoming electron beam axis, an identically zero cross section was theoretically predicted, at odds with the measurement.

In this talk we discuss this interesting geometry and show that the discrepancy between theory and experiment may be traced to the 'unusual' or 'opposite' parity partial waves that arise when coupling to an initial p state. Calculations that include such partial waves do produce a non-zero triple differential cross section, in reasonable agreement with the measurement. We use a time-dependent close-coupling approach [5], including a partial-wave expansion over the coupled wavefunction, to compute such angular differential cross sections and show examples for Mg and Na.

The Los Alamos National Laboratory is operated by Los Alamos National Security, LLC for the National Nuclear Security Administration of the U.S. Department of Energy under Contract No. DE-AC5206NA25396.

References

[1] K. L. Nixon and A. J. Murray 2011 *Phys. Rev. Letts.* **106** 123201

[2] K. L. Nixon and A. J. Murray 2014 *Phys. Rev. Letts.* **112** 023202

[3] A. D. Stauffer 2014 *Phys. Rev. A* **89** 032710

[4] S. Amami, A. Murray, A. Stauffer, K. L. Nixon, G. S. J. Armstrong, J. Colgan and D. Madison 2014 *Phys. Rev. A* **90** 062707

[5] G. S. J. Armstrong, J. Colgan and M. S. Pindzola 2013 *Phys. Rev. A* **88** 042713

[1]E-mail: jcolgan@lanl.gov

e-I7

Sturmian approach for ionization processes: applications and perspectives

L U Ancarani[*] [1], **G Gasaneo**[†,‡], **J A del Punta**[*,†], **D M Mitnik** [§,‡]

[*] Équipe TMS, SRSMC UMR 7565, Université de Lorraine, 57078 Metz, France
[†] Departamento de Física, Universidad Nacional del Sur, 8000 Bahía Blanca, Argentina
[‡] Consejo Nacional de Investigaciones Científicas y Técnicas (CONICET), Argentina
[§] Instituto de Astronomía y Física del Espacio (IAFE) and Universidad de Buenos Aires, Argentina

Synopsis The Sturmian approach finds applications in the description of ionization processes of atoms and molecules. The present developments and some perspectives are discussed.

In the quantum description of ionizing processes of atoms and molecules, whether by impact of photons or charged particles, one needs to deal in the final channel with n charged particles in the continuum. The three-body case (say two electrons and a heavy charged nucleus) has been investigated over the last decades and is the key to many scattering studies.

One important theoretical issue is how to impose to the scattering wave function the correct asymptotic behavior. In many methods the problem is solved using two-body basis (spherical) functions that generally do not possess the correct behavior at large distances. One exception is given by the Sturmian approach using Generalized Sturmian Functions (GSF) [1]. In this spectral method, the two-body basis functions solve a Sturm-Liouville problem. They may be defined in order to take into account part of the interactions of the problem under consideration, and all behave according to a selected (e.g., outgoing) asymptotic behavior. Three-body scattering wave functions can be built with angularly coupled expansions; the intrinsic properties of the basis set make the GSF method computationally efficient. It has now been applied successfully to $(e, 3e)$ [2] and $(\gamma, 2e)$ [3] on Helium (two-body GSF have been applied also to single photoionization of atoms and molecules [4]). To further improve our knowledge of the double continuum and how the asymptotic regime builds up, we are further developing the Sturmian approach along two other lines of investigations:

(1) In the spatial domain where the three charged particles are far from each other (Ω_0 region) we have the known formal asymptotic behavior, given by Peterkop, expressed in terms of an hyperradius $\rho = \sqrt{r_1^2 + r_2^2}$ and five hyperangular coordinates. To adequately deal with such behavior, we have proposed Hyperspherical GSF, their efficiency being demonstrated - for the moment - only for model S-wave problems [5].

(2) A second line goes with an alternative set of basis functions, named Quasi Sturmian Functions (QSF) [6] which have the advantage of being known in closed form for the case of a Coulomb potential. Like GSF, two-body QSF satisfy a non-homogeneous differential equation, and may be constructed with a selected asymptotic behavior. They provide a superior convergence rate when solving a two-body scattering problem [6]. For the three-body case, we propose a representation using hyperspherical coordinates. While the angular variables are treated in a parametric way, the hyperradial part is obtained generalizing the two-body problem. As a consequence, analytical expressions can be given for these new QSF and the desired Coulomb asymptotic behavior in the hyperradial coordinate can be imposed.

An update of the recent developments and a comparison of the different Sturmian basis functions will be presented at the conference.

We acknowledge the CNRS (PICS project N. 06304) and CONICET (project N. DI 158114) for funding our French-Argentinian collaboration.

References

[1] G. Gasaneo *et al* 2013 *Adv. Quantum Chem.* **57** 153
[2] M. J. Ambrosio *et al* 2015 *J. Phys. B* **48** 055204; M. J. Ambrosio *et al* 2015 *Phys. Rev. A* **91** 012704
[3] J. M. Randazzo *et al* 2015, submitted to *Eur. J. Phys. D*
[4] C. M. Granados–Castro *et al* 2015 *J. Phys: Conf. Series* **601** 012009; C. M. Granados–Castro *et al* 2015, to be published in *Adv. Quantum Chem.*
[5] D. M. Mitnik *et al* 2013 *J. Phys. B* **46** 015202; G. Gasaneo *et al* 2013 *Phys. Rev. A* **87** 042707
[6] J. A. Del Punta *et al* 2014 *J. Math. Phys.* **55** 052101

[1] E-mail: ugo.ancarani@univ-lorraine.fr

e-I8

Fully Differential Study of Coherence and Interference Effects in Ionization of H_2 by Proton Impact

M. Schulz*[,1], S. Sharma*, T.P. Arthanayaka*, A. Hasan*[,†], B.R. Lamichhane*, J. Remolina*, S. Gurung*, S. Akula*, D.H. Madison*

* Physics Department, Missouri University of Science & Technology, Rolla, MO 65409, USA

† Department of Physics, UAE University, Al Ain, Abu Dhabi, UAE

In fully differential cross sections for ionization of simple atoms by ion impact, puzzling discrepancies between experiment and theory were found, which were vividly debated for more than a decade. Recently, an important step towards resolving this puzzle was made when ionization of H_2 by proton impact was studied [1]. Here, indistinguishable diffraction of the projectile from the atomic centers of the molecule can lead to interference structures in measured cross sections. In [1] we demonstrated that this interference structure is only observable if the projectile wave packet is broad enough to illuminate both atomic centers simultaneously, i.e. if the projectile beam is coherent. Later, it was demonstrated that the discrepancies for ionization of atomic targets are to a large extent also due to a lack of projectile coherence [2].

Here, we report a more detailed investigation of projectile coherence and interference effects by performing a kinematically complete coincidence experiment. A 75 keV proton beam was collimated with a slit placed at two different distances from the target and intersected with a very cold H_2 beam. For the large slit distance the beam was coherent and incoherent for the small distance. The recoiling H_2^+ ions were momentum-analyzed using a COLTRIMS apparatus. The deflected projectiles were energy-analyzed using a parallel-plate analyzer and detected by a two-dimensional position-sensitive detector. From the position information the azimuthal and polar projectile angles were obtained so that the projectile momentum could be determined. The electron momentum was obtained from momentum conservation. From this information fully differential cross sections (FDCS) could be extracted.

In Fig. 1 the ratios between the coherent and incoherent FDCS, which represents the interference term, are shown for fixed recoil momentum (0.2 a.u.), polar electron angle θ_{el} (35°), and energy loss ε (30 eV) as a function of the azimuthal electron ejection angle ϕ_{el}. The data clearly show that the FDCS for a coherent and incoherent beam differ substantially.

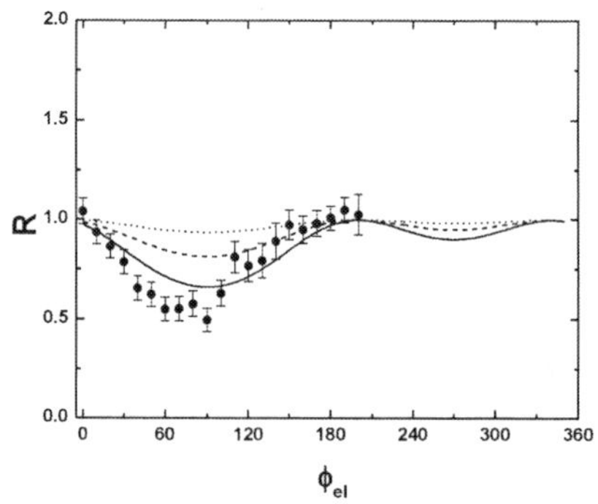

Figure 1. Fully differential interference term for ε = 30 eV and fixed recoil-ion momentum.

The phase angle for two-center molecular interference was assumed to be determined by the recoil ion momentum. However, in Fig. 1 the interference term is plotted for fixed recoil momentum and yet a pronounced structure in R is observed. This suggests that the data are not dominated by molecular two-center, but by some other type of interference. One possibility is interference between different impact parameters leading to the same scattering angle, to which we refer as single-center interference.

The phase angle in single-center interference is expected to be given by the momentum transfer and the average impact parameter separation Δb between the interfering transition amplitudes. The solid curve in Fig. 1 shows the single-center interference term assuming Δb = 2 a.u. The good agreement of the data with this curve leads us to conclude that for ε = 30 eV single-center interference is more important than molecular two-center interference. In contrast, at ε = 57 eV we find two-center interference to be important as well.

This work was supported by the National Science Foundation.

References

[1] K.N. Egodapitiya *et al* 2011 *Phys. Rev. Lett.* **106** 153202

[2] X. Wang *et al.* 2012 *J. Phys. B* **45** 211001

"Double-slit" interferences observed in dielectronic transitions in collisions between hydrogen molecular ion and helium atom

S. F. Zhang *† 1, **D. Fischer** †, **M. Schulz** †‡, **A.B. Voitkiv** †, **A. Senftleben** †, **A. Dorn** †, **J. Ullrich** †II, **X. Ma** * and **R. Moshammer** †

* Institute of Modern Physics, Chinese Academy of Sciences, 730000 Lanzhou, China
†Max-Planck-Institute for Nuclear Physics, Saupfercheckweg 1, D-69117 Heidelberg, Germany
‡Physics Department and LAMOR, Missouri University of Science and Technology, Rolla, Missouri 65409, USA

IIPhysikalisch-Technische Bundesanstalt, Bundesallee 100, D-38116, Braunschweig, Germany

Synopsis Double-slit interferences were studied in collisions between the hydrogen molecular ion and the helium atom where the hydrogen molecular ion played the role of double slits when the helium atomic matter wave scattered on them. In the process under investigation the helium atom was singly ionized and simultaneously the molecular hydrogen ion was dissociated. Different collision mechanisms are identified and interference fringes emerging from a correlated first-order mechanism and from an independent second-order process were observed.

Young's double-slit interference is a clear manifestation of the wave character of light. However, these phenomena are difficult to observe with massive particles. In 1960 Tuan and Gerjuoy discussed the feasibility of an atomic version of Young's double slit experiment in ion-atom collisions [1]. Their theoretical predictions were recently confirmed in capture collision of O^{8+} on D_2 where the capture cross sections vary as a function of the deuterium molecular orientation [2]. Later, interference effects in electron emission from H_2 colliding with 60 MeV/u Kr^{34+} were also observed [3]. The first direct observation of double-slit interference fringes, analogous to that in the optical experiment, was reported by Schimidt *et al* in 2008 [4].

In this symposium, we report the first observation of double-slit fringes in the transverse momentum transfer spectra in collisions of H_2^+ on He (see Figure 1) [5]. During the collision, the target helium atom was singly ionized, meanwhile the hydrogen molecular ion underwent a transition from the ground state $1s\sigma_g$ to the first dissociative state $2p\sigma_u$. There, the hydrogen molecular ion played the role of double slits when the helium atomic matter wave scattered on them.

In this reaction channel, dielectronic transitions could be induced either by the correlated first-order process (through the electron-electron interaction), or by the uncorrelated second-order procedures (through two sequent electron-nucleus interactions). These two mechanisms can be separated experimentally which allows the presentation of different double-slit interference patterns.

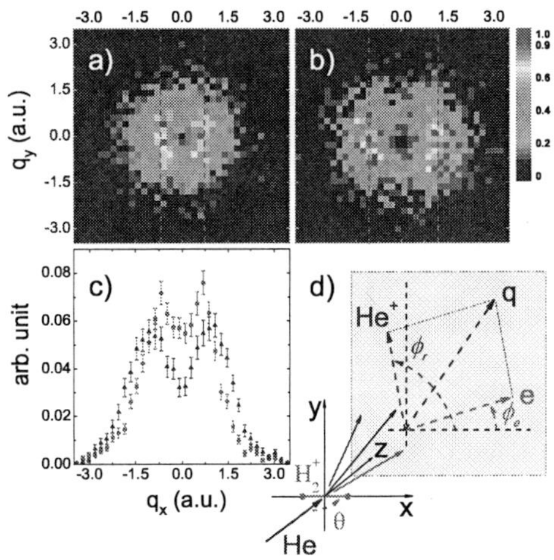

Figure 1. Two-dimensional momentum transfer distributions in the transverse plane under different conditions of internuclear distances. Molecular axis perpendicular to the beam direction: a) *KER* $< 2.5\ eV$; b) *KER* $> 8\ eV$; c) their q_x distributions: • for a), and ▲ for b). d) sketch of the molecular frame and corresponding definitions.

[1]E-mail: zhangshf@impcas.ac.cn

References

[1] T. F. Tuan and E. Gerjuoy, 1960 *Phys. Rev. Lett.* **117** 756

[2] S. Chen *at al* 1993 *Phys. Rev. A* **47** 3923

[3] N. Stolterfoht *at al* 2001 *Phys. Rev. Lett.* **87** 023201

[4] L. P. H. Schmidt *at al* 2008 *Phys. Rev. Lett.* **101** 173202

[5] S. F. Zhang *at al* 2014 *Phys. Rev. Lett.* **112** 023201

e-I10

Towards making the molecular orbital movies by time-resolved (e,2e) electron momentum spectroscopy

Masakazu Yamazaki[1] **and Masahiko Takahashi**[2]

Institute of Multidisciplinary Research for Advanced Materials, Tohoku University, Sendai 980-8577, Japan

Synopsis We report the current status of the time-resolved (e,2e) electron momentum spectroscopy. As a new experimental example, the results of the toluene molecule in the $S_1(\pi, \pi^*)$ state is presented to demonstrate that the technique selectively probes ionization transition to two-hole-one-particle configurations that are hardly accessible from the ground electronic state. This observation demonstrates the potential ability of the present technique to probe ionization of not only outermost orbitals but also of all other, more tightly bound orbitals in molecular excited states.

Through many years of intensive studies, (e,2e) electron momentum spectroscopy (EMS) has been established as a means to measure molecular orbital patterns in momentum space [1,2]. Since the change of the behaviour of electrons in molecular orbital is the driving force behind any reactions, application of EMS to transient species is expected to provide new insights into ultrafast molecular dynamics. For this purpose, we have recently developed time-resolved EMS (TREMS) [3] by combining EMS with the pump-probe technique using an ultrashort pulsed electron beam. The experimental feasibility of orbital imaging for molecular excited states has been successfully proved for the outermost orbital of the S_2 state of deuterated acetone with a lifetime of 13.5 ps [4]. Here, in order to further demonstrate the unique ability of TREMS, we have applied TREMS to a long-lived species, the $S_1(\pi, \pi^*)$ state of toluene with a lifetime of ~80 ns.

Briefly, output from a 5-kHz Ti: sapphire femtosecond laser (<120 fs) was frequency tripled (267 nm) and was then split into a pump path and an electron-generation path. The pump laser (16 μJ, 2.5 kHz) was introduced into the vacuum chamber to excite toluene molecules. On the other hand, the 5-kHz electron-generation laser was directed toward a back-illuminated photocathode (40-nm Ag film) to generate a train of 4 or 8 ultrashort electron pulses (~1 ps) with 0.3 or 0.7 ns intervals that induced EMS reactions of the excited molecules.

Fig. 1 shows a TREMS electron binding energy spectrum of the S_1 toluene. While the data statistics leaves much to be desired, one may see the ionization bands at around 4.5 eV and 17 eV, which can be attributed to the ionization from the S_1 state. The theoretical calculations using the symmetry-adapted cluster configuration interaction method support this observation, showing that the final ionic states are dominated by the two-hole-one-particle configurations in which the electron excitation from the π to π* orbital is involved. Although there is ample room for improvements, mainly in data statistics and energy resolution, present results demonstrate clearly the basic ability of TREMS to enable one to look at arbitrary molecular orbitals in excited states.

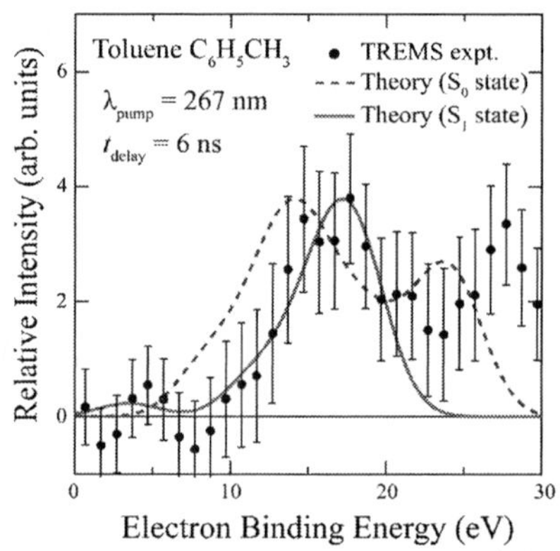

Figure 1. Comparison of TREMS binding energy spectra between experiment and theory for the S_1 excited state of toluene.

References

[1] E. Weigold and I. E. McCarthy 1999 *Electron Momentum Spectroscopy* (New York: Kluwer)

[2] M. Takahashi 2009 *Bull. Chem. Soc. Jpn.* **82** 751

[3] M. Yamazaki *et al* 2013 *Rev. Sci. Instrum.* **84** 063105

[4] M. Yamazaki *et al* 2015 *Phys. Rev. Lett.* **114** 103005

[1] E-mail: masakazu@tagen.tohoku.ac.jp

[2] E-mail: masahiko@tagen.tohoku.ac.jp

Ionization of Laser-Excited Atoms: the Shape of the Cross Sections

A. D. Stauffer

Dept. of Physics & Astronomy, York University, Toronto, Canada

Synopsis We analyze the shape of the cross section for the ionization of laser-excited atoms as a function of the angle of the linear polarization of the laser radiation.

There have been recent experiments in Mg [1, 2] where the target atom was first excited from the ground state by linearly-polarized laser radiation and then ionized by electron impact. The ground $(3s)^2$ configuration is a spherically symmetric state with total angular momentum J = 0 while the excited state with configuration 3s3p has $J = 1$. Since the excitation process produces atoms aligned along the direction of polarization of the radiation, the triple differential cross section will be a function of this direction. Using simple rotation matrices the general shape seen in the measurements in the case where the laser beam is perpendicular to the scattering plane can be derived.

The case where the laser beam is along the direction of the incident electron beam can be analyzed in a similar manner but the results do not agree well with the measurements. Reasons for this discrepancy will be discussed.

The methods used in the above derivation are general and depend only on the angle of alignment of the linear polarization of the laser and the total angular momentum of the excited state [3, 4]. There is no assumption about the method used to evaluate the cross sections for ionization. A recent distorted wave calculation of these processes in Mg [5] confirms the details of the above analysis.

[1] K. L. Nixon and A. J. Murray, Phys. Rev. Lett. **106**, 123201 (2011).

[2] K. L. Nixon and A. J. Murray, Phys. Rev. Lett. **112**, 023202 (2014).

[3] A. D. Stauffer, Phys. Rev A **89** 032710 (2014)

[4] A. D. Stauffer, Phys. Rev A **89** 049906 (2014)

[5] Sadek Amami, Kate Nixon, Gregory Armstrong , Andrew Murray, James Colgan , Al Stauffer and Don Madison, Phys. Rev A **90** 062707 (2014)

E-mail: stauffer@yorku.ca

e-I12

Interaction of twisted light with atomic and molecular targets

Andrey Surzhykov[*1], **Anton Peshkov**[*], **Valery Serbo**[†], **Stephan Fritzsche**[*,‡]

[*] Helmholtz–Institut Jena, D–07743 Jena, Germany
[†] Novosibirsk State University, RUS–630090, Novosibirsk, Russia
[‡] Friedrich–Schiller–Universität Jena, D–07743 Jena, Germany

Synopsis The recent theoretical progress in the analysis of interaction of twisted light with matter is reviewed. Special attention is paid to the excitation of many–electron atoms and ionization of diatomic molecules by the incident Bessel photons. We show, for this case, that the population of excited atomic states and angular distribution of photo–electrons differ significantly from what is expected for the incident plane–wave radiation.

The beams of photons carrying a non–zero projection of the orbital angular momentum (OAM) upon their propagation direction have been in the focus of research over the last years. These *twisted* (or vortex) light beams are produced with the help of spiral phase plates, computer–generated holograms and helical undulators, and used to explore the role of the OAM in the coupling between radiation and matter. In the past, for example, a number of studies have dealt with the OAM transfer to colloidal particles and Bose–Einstein condensates. Moreover, the interaction of vortex photons with *single* atoms and molecules has attarcted much recent attention. These investigations aim to better understand the orbital momentum effects in *fundamental* processes such as the photo–absorption and ionization, as well as Compton and Thomson scattering.

In this contribution, we summarize the recent advances in the theoretical description of interaction of atomic and molecular targets with the so–called Bessel twisted photons. The Bessel wave propagates along the quantization (z–) axis with the well–defined longitudinal momentum k_z and the projection of the total angular momentum m. It is characterized by the vector potential:

$$\mathbf{A}_{\kappa m k_z \lambda}(\boldsymbol{r}) = \int \boldsymbol{e}_{\boldsymbol{k}\lambda}\, \mathrm{e}^{i\boldsymbol{k}\boldsymbol{r}}\, a_{\kappa m}(\boldsymbol{k}_\perp)\, \frac{d^2 k_\perp}{(2\pi)^2} \qquad (1)$$

where $\kappa = |\boldsymbol{k}_\perp|$ is the absolute value of the transverse momentum, and the amplitude $a_{\kappa m}(\boldsymbol{k}_\perp) = (-i)^m\, \mathrm{e}^{im\phi}\, \sqrt{2\pi/\kappa}\, \delta(k_\perp - \kappa)$. With the help of the potential (1) we have recently studied the absorption of vortex light by many–electron atoms [1]. It was shown, in particular, that the transition selection rules and, hence, the magnetic population of excited atomic sub–levels can be strongly affected by the OAM of the incident radiation. Based on our calculations, we argue that the OAM–effects can be easily observed by measuring the linear polarization of the subsequent fluorescent emission.

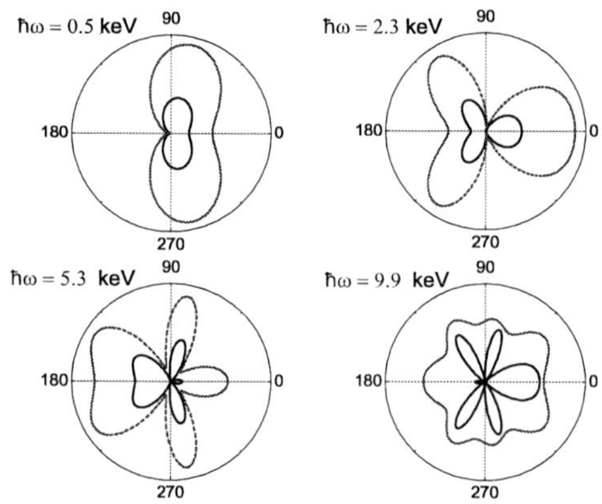

Figure 1. Angular distribution of electrons ionized from the hydrogen molecular ion H_2^+ by the plane–wave (solid line) and twisted (dashed line) photons. The molecule is aligned along the axis, tilted by the angle 45 deg with respect to the incident light direction.

Beside the photo–excitation of atoms, the ionization of diatomic molecules by the Bessel radiation has been investigated. For this case, we found that the Young–type interference behaviour of the angular distribution of emitted electrons is also influenced by the OAM and the kinematic parameters of the beam (1). As one can see from Fig. 1, for example, the electron emission pattern predicted for the incident twisted light differs a lot from what is expected for the plane–wave radiation.

References

[1] A. Surzhykov *at al* 2015 *Phys. Rev. A* **91** 013403

[1]E-mail: a.surzhykov@gsi.de

Ion-momentum imaging of dissociative-electron-attachment dynamics in CO_2, N_2O, HCCH and CF_4

M. Fogle*[1], A.L. Landers*, A. Moradmand*, J. Sartor*, D.S. Slaughter†, D.J. Haxton†, T.N. Rescigno†, C.W. McCurdy†⋆, Th. Weber†, A. Belkacem†, A.E. Orel⋆, S. Matsika‡

* Department of Physics, Auburn University, Auburn, Alabama 36849, USA
† Lawrence Berkeley National Laboratory, Chemical Sciences, Berkeley, California 94720, USA
⋆ Department of Chemistry and Applied Science, University of California, Davis, California 95616, USA
‡Department of Physics, Temple University, Philadelphia, Pennsylvania 19122, USA

Synopsis We present experimental results for dissociative electron attachment to CO_2, N_2O, HCCH and CF_4 from an ion-momentum imaging technique. From our measured ion-momentum results we extract fragment kinetic energies and angular distributions. We directly observe the dissociation dynamics associated with the formation of transitory negative ions. We compare the experimental results with *ab initio* electronic structure and fixed-nuclei scattering calculations and obtain good agreement.

In recent years, low energy dissociative electron attachment (DEA) interactions have been of interest to varying biological and technological applications. To study the dynamics resulting from DEA, we used an ion-momentum imaging apparatus based on the Cold Target Recoil Ion Momentum Spectroscopy (COLTRIMS) technique in which a molecular beam is crossed by a pulsed electron beam. The beam interaction takes place in a 4π pulsed electrostatic spectrometer that collects the anion fragments resulting from DEA. The molecular beam is formed by a supersonic expansion which results in a well-localized and cold target. An overview of the apparatus is given in Ref. [1].

Using this apparatus we have investigated the DEA dynamics for several small molecules: CO_2 at the 4 eV shape resonance [2] and the 8 eV Feshbach resonance [3]; N_2O at the 2.3 eV shape resonance [4]; HCCH at the 3 eV shape resonance [5]; and CF_4 near the 7 eV resonance.

The experimental ion-momentum results are compared to *ab initio* electronic structure and fixed-nuclei scattering calculations to gauged the resulting dynamics driven by DEA. In many cases, conical intersections play a pivotal role in driving the dynamics. Some of these systems exhibit non-axial recoil conditions indicative of a bending dynamics in the transitory negative ion state while others exhibit a direct axial recoil dissociation without any bending.

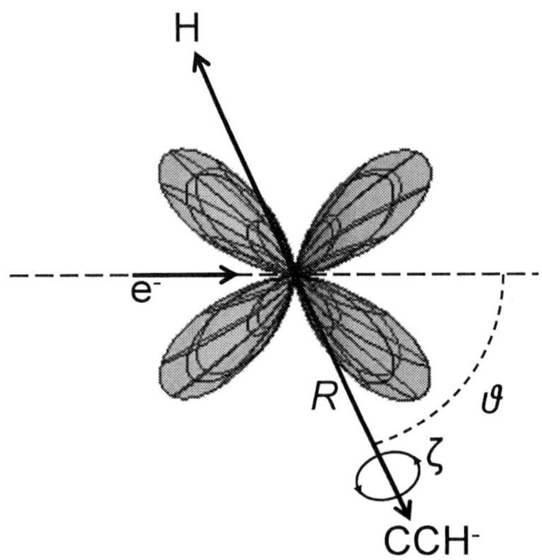

Figure 1. Modified axial recoil approximation for DEA to HCCH. The attachment amplitude lobes are shown along with the incoming electron and outgoing dissociation fragments in a rotated frame.

References

[1] A. Moradmand, J. Williams, A.L. Landers, M. Fogle 2013 *Rev. Sci. Instrum.* **84**, 033104

[2] A. Moradmand, D.S. Slaughter, A.L. Landers, M. Fogle 2013 *Phys. Rev. A* **88**, 022711

[3] A. Moradmand, D.S. Slaughter, D.J. Haxton, A.L. Landers, C.W. McCurdy, T.N. Rescigno, A. Belkacem, M. Fogle 2013 *Phys. Rev. A* **88**, 032703

[4] A. Moradmand, A.L. Landers, M. Fogle 2013 *Phys. Rev. A* **88**, 012713

[5] M. Fogle, D.J. Haxton, A.L. Landers, A.E. Orel, T.N. Rescigno 2014 *Phys. Rev. A* **90**, 042712

[1]E-mail: fogle@physics.auburn.edu

Coherent emission from diatomic molecules: from femto- to atto-seconds

Omar Ariel Fojón[*1]

[*] Instituto de Física Rosario (CONICET-UNR), Bvd. 27 de Febrero 210 bis, (2000) Rosario, Argentina

Synopsis We present the state of art of our work on interference effects observed in the ionized electron spectra from diatomic molecules [1] by impact of energetic heavy ions, electrons and photons. We put in evidence interference effects coming from the coherent emission from both molecular centers in a variety of scenarios where the interaction times of the projectile with the target ranges from some femto-seconds to the sub-femtosecond regime. In the latter case, we present recent results for photoionization by trains of attosecond pulses assisted by near infrared lasers [2]. Under these conditions, we observe not only the mentioned two-center interferences and confinements effects [3, 4] but also satellite interferences coming from the presence of the laser field.

Interference effects coming from the coherent emission from diatomic molecules was first recognized by Cohen and Fano in theoretical studies of photoionization of N_2 and O_2 [5]. They showed that electrons can be emitted coherently from both molecular centers giving place to interferences in the electron spectra equivalent to that seen in macroscopic double-slit experiments. However, the first experimental clear proof for these interference phenomena was obtained for the single ionization of H_2 molecules by impact of Kr^{34+} projectiles. After that, coherent electron emission from diatomic molecules was studied not only for heavy ions but also for electrons [6] and photons [3, 4]. In all these cases, the effective interaction time is in the femtosecond or sub-femtosecond domain. In 2001, the advent of attopulses [7] paved the way to the attophysics field where the typical experiment is the pump-and-probe schema. For instance, a molecule is ionized by an extreme ultraviolet (XUV) attopulse (pump) and then the photoelectorn spectra is modulated (probe) by a near infrared laser (NIR). This in turn opened the way to new coherence effects coming from interferences between the attopulse and the NIR. By using a separable Coulomb-Volkov model, we study the photoelectron angular distribution (PAD) from H_2^+ for polarization parallel to the internuclear axis. In Fig. 1, we observe assymetries *up* (a) and *down* (c) produced by the interferences among the different n channels (b) and (d), respectively, opened by the NIR and provoked by variation of the delay t_0 between the XUV attopulse and the NIR [2]. This is a simple example showing that pump-and-probe schemes are a promising tool to reach accurate control of electron processes in atomic and molecular samples in real time.

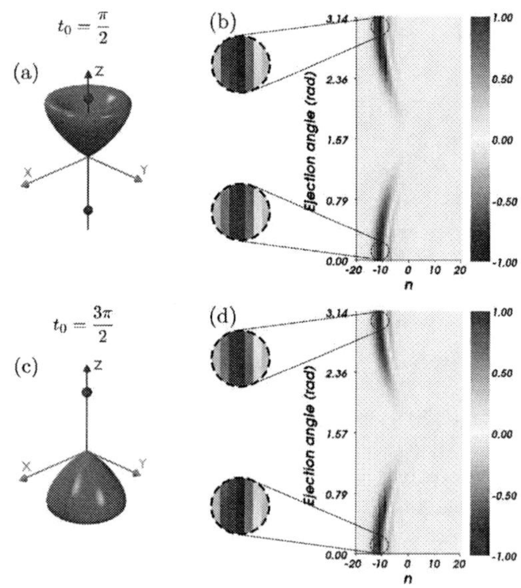

Figure 1. PAD for H_2^+ by a XUV attopulse (energy 5 a.u., duration time 200 as) assisted by a NIR laser(800 nm, 10^{-12} W/cm^2) [2]. See text.

[1] E-mail: fojon@ifir-conicet.gov.ar

References

[1] M. F. Ciappina, O. A. Fojón, and R. D. Rivarola 2014 Topical Review: *J. Phys. B: At. Mol. Opt. Phys.* **47** 042001.

[2] D.I.R. Boll and O. A. Fojón 2014 *Phys. Rev. A* **90** 053414.

[3] J. Fernández, O. Fojón, A. Palacios, and F. Martín 2007 *Phys. Rev. Lett.* **98** 043005.

[4] J. Fernández, O. Fojón, and F. Martín 2009 *Phys. Rev. A* **79** 023420.

[5] C. Cohen and U. Fano 1966 *Phys. Rev.* **150** 30.

[6] C. R. Stia, O. A. Fojón, P. Weck, J. Hanssen and R. D. Rivarola 2003 *J. Phys. B: At. Mol. Opt. Phys.* **36** L257.

[7] M. Hentschel *et al* 2001 *Nature* **414** 509.

Accuracy of theory for calculating 3-Body and 4-Body fully differential cross sections for electron-impact ionization of atoms and molecules

Don Madison* [1]

*Department of Physics, Missouri University of Science and technology, Rolla, MO, USA

Synopsis This talk will show the current status of agreement between experiment and theory for electron impact ionization of atoms and molecules. Results will be presented for both 3-body (1 active electron) and 4-body processes (2 active electrons) for cases in which the orientation of the target is not known and cases for which the orientation of the target is known.

In the last two decades, there have been numerous studies of high-energy electron impact single ionization of atoms and molecules. These studies were called EMS (Electron Momentum Spectroscopy) and they were very valuable in determining the accuracy of target wavefunctions since the measured cross sections were proportional to the momentum space wavefunction of the ejected electron.

Lower energy collisions are much more difficult for theory since the detailed kinematics of the experiment become important. For ionization of atomic hydrogen and helium, essentially exact close-coupling calculations (convergent close coupling (CCC) [1], exterior complex scaling (ECS) [2], and time dependent close coupling (TDCC) [3]) yield excellent agreement with experimental measurements. For larger atoms, agreement between experiment and theory is not nearly as good. Very recently, an extensive set of measurements both inplane and out-of-plane for 64 eV electron impact ionization of Ne were compared with various theoretical calculations and it was found that both the BSR (B-spline R-matrix with pseudostates) and the 3DW (3-body distorted wave) approach gave very good agreement with relatively absolute experimental data [4].

For ionization of molecules, the most commonly used theoretical methods are the first Born approximation (FBA) [5], the FBA two-center continuum (FBA-TCC) approximation with correct boundary conditions in the entrance and exit channels [6], the molecular 3-body distorted wave (M3DW) approximation [7,8], and the time dependent close coupling (TDCC) approximation [9].

[1] E-mail: madison@mst.edu

The FBA and FBA-TCC are high energy approximations and the FBA-TCC works reasonably well for high energies. The TDCC has been developed only for H_2 so far and it works reasonably well for low energy single ionization of H_2. One of the strengths of the M3DW approximation is that it can be applied for any energy, any size molecule and both 3-body (one active electron) and 4-body (two active electrons) processes.

In this talk we will show the current status of agreement between experiment and theory for low and intermediate energy electron-impact single ionization of atoms and molecules for both 3-body and 4-body processes. Results will be shown for ionization of atoms and molecules when the orientation of the target is not known, as well as ionization of atoms and molecules when the orientation of the target is known.

References

[1] I. Bray and D.V. Fursa 1996, *Phys. Rev. A* **54**, 2991
[2] T. N. Rescigno *et al* 1999, *Science 286,* 2474
[3] J. Colgan *et al* 2002, *Phys. Rev. A 65,* 042721
[4] X. Ren *et al* 2015, Phys. Rev. A **91**, 032707
[5] C. Champion *et al* 2002 *J. Chem. Phys.* **117** 197
[6] P. Weck *et al* 2002 Phys. Rev. A **66** 012711
[7] D.H. Madison and O. Al-Hagan 2010 J. Atomic, Molecular, and Optical Physics 367180
[8] I Toth and L Nagy 2010 J. Phys. B. **43** 135204
[9] J. Colgan *et al* 2008 Phys. Rev. Lett. **101** 233201

e-I16

Three-body fragmentation of simple molecules induced by electron impact multiple ionizations

Xiangjun Chen*,† [1], Enliang Wang*,†, Xu Shan*,†, Zhenjie Shen*,†, Maomao Gong*,†, Yaguo Tang*,†

* Heifei National Laboratory for Physical Sciences at Microscale and Department of Modern Physics, University of Science and Technology of China, Hefei, Anhui, 230026, China

† Synergetic Innovation Center of Quantum Information and Quantum Physics, University of Science and Technology of China, Hefei, Anhui, 230026, China

Synopsis Three-body fragmentation of simple molecules induced by electron impact multiple ionizations has been investigated. The dissociation mechanisms are clearly distinguished by the combined analysis using Dalitz plots together with momentum correlations.

Taking the advantages of the rapidly developing momentum imaging technique, it is possible for the experimentalists to explore the fragmentation dynamics of molecules [1, 2]. In these related studies, an interesting question is under what conditions the chemical bonds of multiply ionized polyatomic molecules break through non-sequential or sequential processes. Recently, the three-body fragmentation has captured continuous interest in the studies of dissociative ionizations by collisions of swift [3, 4] and slow heavy ions [5], x-rays [6], as well as intense laser fields [7]. However, the investigation of many-body fragmentation dynamics of the multiply ionized molecule by electron collision is still very difficult due to the extremely low ionization cross sections [8].

In this talk, the investigations of three-body fragmentation processes of CO_2^{q+} and OCS^{q+} (q =3, 4) induced by electron collision at an impact energy of 500 eV will be presented. A high-efficiency multi-coincidence technique, which largely improves the coincident count rate, makes it possible for us to separately detect the sequential and non-sequential fragmentation processes. The dissociation mechanism is clearly distinguished by the combined analysis using Daliz plots and momentum correlations.

The total Newton diagrams for three-body fragmentations of CO_2^{3+} and OCS^{3+} are shown in Figure 1 (a) and (b). The momentum vector of one O^+ in CO_2^{3+} and S^+ in OCS^{3+} are fixed along x axis while the momenta of another two fragment ions are located in the lower and upper halves of the plots. As shown in the figure, the two intense islands represent the linear or molecular bending dissociations. Each of the islands drags a half-circle tail, which is the evidence of the sequential fragmentations.

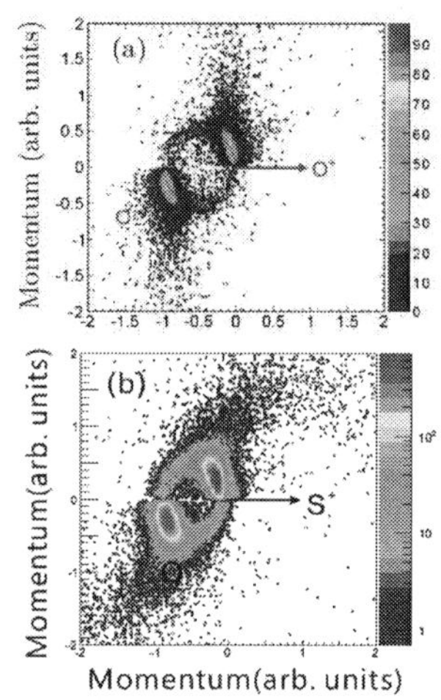

Figure 1. Newton diagrams for (a) three-body fragmentation of CO_2^{3+} and (b) three-body fragmentation of OCS^{3+}.

References

[1] R. Dörner *et al* 2000 *Phys. Rep.* **330** 95.
[2] J. Ullrich *et al* 2003 *Rep. Prog. Phys.* **66** 1463.
[3] L. Adoui *et al* 2006 *Nucl. Instrum. Meth. B* **245** 94.
[4] M. R. Jana *et al* 2011 *Phys. Rev. A* **84** 062715.
[5] N. Neumann *et al* 2010 *Phys. Rev. Lett.* **104** 103201.
[6] R. K. Singh *et al* 2006 *Phys. Rev. A* **74** 022708.
[7] C. Wu *et al* 2013 *Phys. Rev. Lett.* **110** 103601.
[8] E. L. Wang *et al* 2015 *Phys. Rev. A* **91** 052711.

[1] E-mail: xjun@ustc.edu.cn

A thorough study of Young-type interferences in (e,2e) on H_2 molecules with known spatial alignment

Alexander Dorn* [1], **Xueguang Ren***,† , **Xingyu Li**+,#, **Khokon Hossen*** and **Xiangjun Chen**+,#

*Max Planck Institute for Nuclear Physics, 69117 Heidelberg, Germany
† Physikalisch-Technische Bundesanstalt, 38116 Braunschweig, Germany
+ Hefei National Laboratory for Physical Sciences at Microscale and Department of Modern Physics, University of Science and Technology of China, Hefei, Anhui, 230026, China
# Synergetic Innovation Center of Quantum Information and Quantum Physics, University of Science and Technology of China, Hefei, Anhui, 230026, China

Synopsis An experimental and theoretical (e,2e) study is performed for H_2 at 500 eV impact energy. The molecular alignment is determined experimentally by measuring the emitted proton momentum for ground state dissociation of the residual ion. We observe oscillations in the cross section as function of the molecular alignment angle which can be reproduced using a Multi-Center Distorted Wave method (MCDW) while the standard treatment using a coherent sum of effective atomic amplitudes fails completely.

Interferences of coherent electron waves emitted from multicenter systems or traveling along several indistinguishable paths are ubiquitous in physics. One prominent example is their observation for electron emission from gas phase molecules induced by photo-absorption or particle impact. Often analogies are drawn to Young's double slit experiment and the simplest models describing molecular ionization are based on this scenario by using single center outgoing waves which are coherently summed up for all emission centers in the molecule. Two center interferences were also considered for the (e,2e) reaction where, so far, all experiments have averaged over the molecular axis alignment in space which is one essential parameter which determines the interference pattern.

Here we present fully differential cross sections (FDCS) for fast (500 eV) electron impact on H_2 with the H_2^+ ($^2\Sigma^+_g$) final ionic ground state. A small fraction of ions dissociates such that the measurement of the proton momentum vector allows to determine the spatial alignment of the molecular axis in space during the collision. Experimentally a reaction microscope is used to detect both outgoing electrons and the proton in a triple coincidence. Theoretically a Multi-Center Distorted Wave method (MCDW) is used and, alternatively, the standard interference factor $(1+\cos(\vec{\chi}\cdot\vec{\rho}))$ is multiplied with the atomic hydrogen FDCS. Here $\vec{\rho}$ is the internuclear vector and $\vec{\chi}$ is effectively the recoil ion momentum. In Figure 1a we see strong alignment dependent variations of the FDCS which are well reproduced by MCDW (b) while the pure interference description (c) fails clearly.

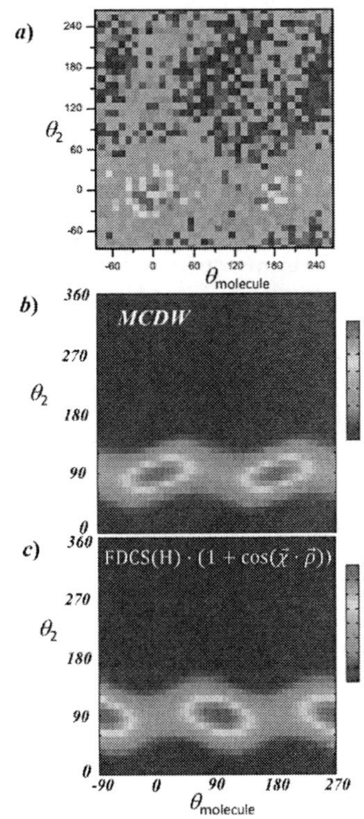

Figure 1. FDCS as function of the molecular alignment angle $\theta_{molecule}$ and the ejected electron angle θ_2 both in the plane perpendicular to the incoming projectile beam ($\theta_2 = 0°$ is roughly the direction along the momentum transfer). Projectile scattering angle $\theta_1 = -20°$, ejected electron energy $E_2 = 10$ eV. a) experiment. b) MCDW calc. c) interference factor.

[1] E-mail: A.Dorn@mpi-k.de

ph-I1

Coincidence photoelectron measurements following 2p photoionization in Mg

Emma Sokell*[1], P Bolognesi†, Stephan Safgren*, Elaine Long* and Lorenzo Avaldi†

• School of Physics, Science Centre North, University College Dublin, Belfield, Dublin 4, Ireland
† CNR-Istituto di Struttura della Materia , Area della Ricerca di Roma1 , 00015 Monterotondo Scalo, Italy

Synopsis Coincidence photoelectron spectroscopy has been used to test the two-step model of Auger decay following 2p photoionization in magnesium. Fitting of the data has yielded information on the dipole photoionization matrix elements as a function of photon energy.

Photoelectron-Auger electron coincident measurements have been used to investigate 2p photoionization in magnesium. This target is interesting because information can be extracted about the dipole matrix elements. Generally, the angular distribution of the Auger electron depends on both the alignment of the core hole state and the contributing partial waves of the Auger decays. In specific cases, such as Mg $2p_{3/2}$ photoionization followed by L_3-M_1M_1 Auger decay, only a single partial wave contributes to the Auger decay and a measurement of the angular distribution of the Auger electron yields information about the alignment of the core-hole state [1]. This particular case led to a complete experiment, in which the dipole matrix elements were determined at a photon energy of 80 eV, without measurement of the spin of the electrons involved [2]. Similar information about the dynamics of the photoionization process [1] can be achieved by coincidence measurements of the photo- and Auger electrons, by a proper choice of the photoelectron angle.

In the present work the $2p_{3/2}$ photoelectron and the L_3-M_1M_1 Auger electron of Mg have been measured in coincidence after angle and energy selection. The experiments were performed with the multi-coincidence end-station of the Gas Phase beamline at Elettra. This set-up contains ten hemispherical analyzers mounted on two independently rotatable frames. Three of the analyzers, in the plane of the linearly polarized incident light, direction ε, were fixed at 0, 30 and 60° with respect to ε, whilst the other seven, also mounted at 30° intervals in the same plane, were rotated in the plane perpendicular to the direction of propagation of the incident beam. This arrangement allows three triple differential cross-section (TDCS) data sets to be obtained simultaneously.

Sets of three TDCS measurements have been made at a range of photon energies. One made at 80 eV, allowed a direct comparison with previous measurements [3]. A simultaneous fit of the three TDCS indicates agreement with the predictions of the two-step model [1] and yields information on the dipole matrix elements at 80 eV. A similar fitting procedure carried out at other photon energies, not only provides information on the dipole matrix elements, which can be compared with theory [4], but shows that the two-step predictions become less valid as the photon energy is reduced and final state interactions, such as post-collision interactions, become significant.

This work was funded by Science Foundation Ireland, grant number SFI (08/RFP/PHY1117).

References

[1] V. Schmidt 1994 *Nucl. Instrum. Meth. in Phys. B.* 87 241.
[2] A. Hausmann *et al* 1988 *Phys. Rev Lett.***61** 2669.
[3] H. Lörch *et al. 1999 J. Phys. B.* **32** L371.
[4] P. Deshmukh and S. Manson 1983 *Phys. Rev. A* **28** 209.

[1]E-mail: emma.sokell@ucd.ie

XFEL-induced ultrafast electron and molecular dynamics

Kiyoshi Ueda* [1]

*Institute of Multidisciplinary Research, Tohoku University, Sendai 980-8577, Japan

Synopsis Current status of our study on hard x-ray-free-electron-laser (XFEL)-induced ultrafast electron and molecular dynamics at Japanese XFEL facility SACLA, will be reported.

In March 2012, SPring-8 Angstrom Compact free electron LAser (SACLA), started user operation in Japan [1,2]. We set up the program to investigate the dynamical behavior of heavy atoms as an isolated atom, in the molecule, and in the cluster, with SACLA.

At 5.5 keV, with the fluence of 50 μJ/μm^2, we could identify that Xe^{n+} with n up to 26 is produced, evidencing occurrence of deep inner-shell ionization and sequential electronic decay cycles repeated multiple times in the heavy atom within the XFEL pulse of ~ 10 fs [3]. Reducing the photon energy to 5 keV, with the same fluence, we could identify occurrence of resonance-enabled x-ray multiple ionization [4].

The results for momentum-resolved multiple ion coincidence study on iodine-contained organic molecules illustrates that the charges are produced in the iodine site by the deep inner-shell ionization and sequential electronic decay cycles and spread over the entire molecule within the FEL pulse duration (<10 fs), leading to Coulomb explosion. The light atoms, however, move significantly, a few hundred pm for hydrogen atoms, before completing the ionization-decay cycles within the pulse duration of 10 fs [5].

The results for electron spectroscopy on argon and xenon clusters illustrate that nanoplasma are formed by the XFEL pulse and continuous thermal emission from the plasma occurs in the time scale of ps [6].

The nanoplasma formation, nanoplasma expansion, and nanoplasma recombination have been probed by XFEL pump – near infrared (NIR) laser probe experiments. Namely, NIR laser causes surface Plasmon resonant heating of the nanoplasma produced by the XFEL irradiation at some time delay. We could probe the Plasmon heating via highly charged ion productions and high-energy thermal emission of the heated plasma as a function of NIR pulse delay [7].

We could also succeed to record x-ray scattering images of giant xenon clusters with the size up to 10^7 atoms [8].

The experiments were performed at SACLA with the approval of JASRI and the program review committee. I am grateful to coauthors of Refs. [3-8] for their invaluable contributions. The study was supported by the X-ray Free Electron Laser Utilization Research Project and the X-ray Free Electron Laser Priority Strategy Program of the Ministry of Education, Culture, Sports, Science and Technology of Japan.

References

[1] T. Ishikawa *et al.*, 2012 *Nature Photon.* **6**, 540
[2] M. Yabashi, H. Tanaka, T. Tanaka et al., 2013 *J. Phys. B: At. Mol. Opt. Phys.* **46**, 164001
[3] H. Fukuzawa, S.-K. Son, K. Motomura et al., 2103 *Phys. Rev. Lett.* **110,** 173005
[4] K. Motomura, H. Fukuzawa, S.-K. Son et al., J. Phys. B: At. Mol. Opt. Phys. 46, 164024 (2013).
[5] K. Motomura, E. Kukk, H. Fukuzawa et al., submitted.
[6] T. Tachibana, Z. Jurek, H. Fukuzawa et al., Scientific reports (in press).
[7] W. Q. Xu, Y. Kumagai, H. Fukuzawa et al., in preparation.
[8] T. Nishiyama, C. Bostedt, K. Nagaya et al., in preparation.

[1] E-mail: ueda@tagen.tohoku.ac.jp

ph-I3

Alignment and Orientation of N_2^+

J.E. Furst[1], T.J. Gay[2], J.R. Machacek[2,3] and R.R. Lucchese[4]

[1]School of Mathematics and Physical Sciences, University of Newcastle, Ourimbah 2258, NSW, Australia
[2]Jorgensen Hall, University of Nebraska, Lincoln, Nebraska 68588-0299 USA
[3]Atomic and Molecular Physics Laboratories, The Australian National University, Canberra ACT 0200, Australia
[4]Department of Chemistry, Texas A & M University, College Station, TX 77843-3255 USA

Synopsis Circularly polarized light was used to excite and ionize molecular nitrogen. The orientation and alignment of the excited ion state were used to determine all three continuum channel angular momentum transfer probabilities and the ratio of parity favored to parity unfavored channels.

We are interested in the dynamics of doubly excited molecular states above the first ionization threshold where dissociation and ionization compete with each other. In the first experiment to measure the orientation, $\mathcal{O}_0^{(1)}$, of an excited N_2 state we recently identified an intermediate non-Rydberg doubly-excited resonance (NRDER) of Π symmetry as a dissociation precursor in N_2 [1]. This demonstrated the usefulness of using orientation to untangle states of different symmetries.

We now report on an experiment where we have used circularly polarized radiation to extend the early work of Poliakoff et al [2] and Guest et al [3]. They used linearly-polarized synchrotron radiation to examine the excitation and ionization of N_2 above the first ionization threshold. They determined the alignment ($\mathcal{A}_0^{(2)}$) of the excited state [4] and from that the ratio of the parity unfavored to parity favored channels, γ.

We were able to determine both the orientation and the alignment of the excited state over a range of energies from 18 eV to 32 eV. This allowed a complete determination of all three continuum channel angular momentum transfer probabilities, $\mathcal{S}_i\ (j_i, j_t)$ as well as the channel ratio [5]. In the high-j limit, the $\mathcal{S}_i$ are related to the alignment and the orientation by

$$\mathcal{S}_{+1} = \tfrac{1}{6}(2 + 5\langle\mathcal{A}_0^{(2)}\rangle - 6\ \langle\mathcal{O}_0^{(1)}\rangle),$$
$$\mathcal{S}_{-1} = \tfrac{1}{6}(2 + 5\langle\mathcal{A}_0^{(2)}\rangle + 6\ \langle\mathcal{O}_0^{(1)}\rangle), \qquad (1)$$
$$\mathcal{S}_0 = \tfrac{1}{3}(1 - 5\langle\mathcal{A}_0^{(2)}\rangle).$$

We will present a review of the analysis of orientation and alignment in molecular systems as well as recent results for the following reaction in N_2

$$\gamma + N_2\left(X^1\Sigma_g^+\right) \rightarrow N_2^+(B^2\Sigma_u^+) + e^-\left(\epsilon\pi_g, \epsilon\sigma_g\right) \rightarrow N_2^+\left(X^2\Sigma_g^+\right) + \gamma\ (391\text{ nm}) \qquad (2)$$

[1] E-mail: john.furst@newcastle.edu.au

These results include the linear and circular polarization of the radiation emitted from N_2^+, as well as the orientation and alignment of these states. This allows us to calculate individual channel probabilities as well as the ratio of the parity unfavored to parity favored channels.

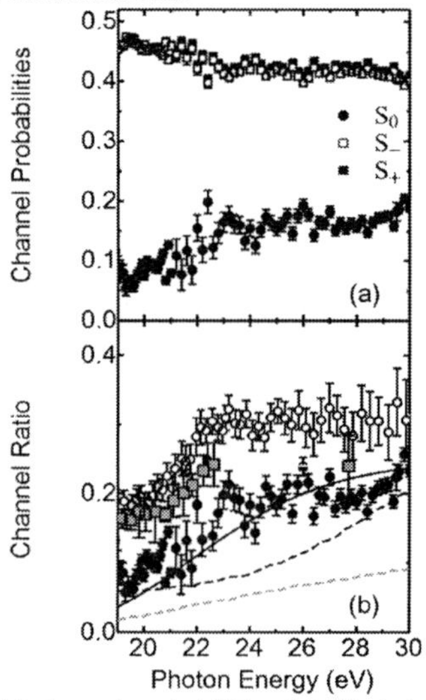

Figure 1: (a) channel probabilities (b) filled circles channel ratio; open circles Guest *et al* [3], our theoretical calculations (solid line), Poliakoff *et al* [2] (green squares), MSM theory [2] (lower dashed line -green), Rescigno *et al.* [6] (upper dashed line -black)

References

[1] J. E. Furst *et al*, Physical Review A **86**, 041401 (2012).
[2] E. D. Poliakoff *et al*, Physical Review Letters **46**, 907 (1981).
[3] J. A. Guest, *et al*, Physical Review A **28**, 2217 (1983).
[4] U. Fano and J. H. Macek, Review of Modern Physics **45**, 553 (1973).
[5] C. H. Greene and R. N. Zare, Annual Review of Physical Chemistry **33**, 119 (1982).
[6] T. Rescigno, *et al*, Journal of Chemical Physics **68**, 970 (1978).

Decoding attosecond electron-nuclear dynamics in molecules by means of XUV-IR and XUV-XUV pump-probe schemes

Alicia Palacios*[1], Alberto González-Castrillo*, Fernando Martín*†

* Departamento de Química, Módulo 13,Universidad Autónoma de Madrid, Cantoblanco 28049, Madrid, Spain
† Instituto Madrileño de Estudios Avanzados (IMDEA) en Nanociencia, Cantoblanco, 28049 Madrid, Spain

Synopsis We theoretically investigate the suitability of XUV-IR and XUV-XUV pump-probe schemes using ultrashort pulses to explore electron dynamics in molecules. We first study the effect of autoionization and nuclear motion in the photoelectron distributions resulting from using a XUV-IR pump-probe protocol. We then demonstrate that by illuminating the molecule with twins XUV attosecond pulses with a given delay one creates an interferometer from which the pumped electronic and vibrational wave packet can be reconstructed.

In the last decade, XUV-IR pump-probe schemes have been employed to explore electron dynamics in photoionization experiments performed with atomic [1, 2] and molecular targets [3, 4]. These works use an attosecond XUV pulse to trigger ionization and the IR field to trace and drive the photoelectron in time and eventually achieve a certain control of the ionization events. Particular interest is focus on the role of autoionization in the laser-induced dynamics [2, 3]. Despite the insight provided by these applications, unraveling and mapping the intrinsic electronic and nuclear dynamics in molecules requires not only the use of short enough laser pulses to launch it, but also probe pulses with wavelengths and intensities that do not modify the intrinsic behavior of the system. We thus propose the use of two XUV pulses of attosecond durations as an ideal pump-probe scheme, because their short wavelengths ensure a negligible distortion of the molecular potential. In this talk, we present our theoretical results in both scenarios: the laser-assisted photoionization of a hydrogen molecule using a XUV-IR pump-probe scheme, with particular emphasis on the electron streaking phenomena [5], and the gentler XUV-XUV pump-probe protocol to obtain a complete characterization, including both amplitudes and relative phases, of a field-free evolving vibronic (electronic and vibrational) wave packet [6].

Firstly, a relatively low intense IR field is used in combination with an XUV attosecond pulse in order to explore the electron streaking patterns. Both, the nuclear motion and the existence of autoionizing states strongly modify the electron streaking patterns from the standard ones observed in direct ionization of atoms. Moreover, the IR field can suppress autoionization from doubly excited states for a range of time delays. We will briefly discuss the state-of-the-art theoretical tools to implement similar approaches for large molecules [4]. Secondly, for the XUV-pump/XUV-probe approach, two twin XUV pulses with a given delay will illuminate the target creating a molecular interferometer. The analysis of the underlying physics is inspired by existing applications using one [7] and two-color schemes [1] to reconstruct electron wave packets from atomic interferometers, where the oscillating patterns observed in the ionization channels result from electrons being ejected through different quantum paths. In our scheme applied to the hydrogen molecule, both direct and sequential two-photon ionization paths leave the molecule in the same final vibronic state. We theoretically demonstrate that the interferometric signal allows for a complete identification of both electronic and nuclear phases in the wave packet generated by the pump pulse. We also show that although total ionization yields reveal entangled electronic and nuclear dynamics in the bound states, doubly differential yields (differential in both electronic and nuclear energies) exhibit in addition the dynamics of autoionization, i.e., of electron correlation in the ionization continuum.

[1]E-mail: alicia.palacios@uam.es

References

[1] T. Remetter *et al* 2006 *Nat. Phys.* **2** 323

[2] S. Gilbertson *et al* 2010 *PRL* **105** 263003

[3] G. Sansone *et al* 2010 *Nature* **465** 763

[4] F. Calegari *et al* 2014 *Science* **346** 336

[5] A. Palacios *et al.* 2014 *J. Phys. B* **47** 124013

[6] A. Palacios *et al* 2014 *Proc. Natl. Acad. Sci.* **111** 3963

[7] J. Feist *et al* 2011 *PRL* **107** 093005

ph-I5

Electron Rearrangement Dynamics in Dissociating Iodine Molecules

K. Schnorr[a,1], G. Schmid[a], S. Augustin[a], A. Senftleben[b], M. Kurka[a], A. Rudenko[c], K. Meyer[a], M. Kübel[d], M.F. Kling[d], Y.H. Jiang[e], B. Siemer[f], M. Wöstmann[f], H. Zacharias[f], R. Mitzner[g], S. Düsterer[h], R. Treusch[h], J. Ullrich[i], T. Pfeifer[a], C.D. Schröter[a], and R. Moshammer[a]

[a]Max-Planck-Institut für Kernphysik, Saupfercheckweg 1, 69117 Heidelberg, Germany
[b]Universität Kassel, 34132 Kassel, Germany
[c]J.R. Macdonald Laboratory, Kansas State University, Manhattan, Kansas 66506, USA
[d]Physics Department, Ludwig-Maximilians-Universität München, 85748 Garching, Germany
[e]Shanghai Advanced Research Institute, Chinese Academy of Sciences, Shanghai 201210, China
[f]Westfälische Wilhelms-Universität, 48419 Münster, Germany
[g]Helmholtz Zentrum Berlin, 14109 Berlin, Germany
[h]Deutsches Elektronen-Synchrotron, 22607 Hamburg, Germany
[i]Physikalisch-Technische Bundesanstalt, 38116 Braunschweig, Germany

Synopsis Electron transfer in dissociating multiply charged iodine molecules (I_2^{n+}) is investigated by XUV-pump–XUV-probe spectroscopy using intense FEL radiation from FLASH. In particular, we determined the critical times and corresponding internuclear distances over which electron transfer takes place.

Electron transfer and migration are driving mechanisms for many fundamental processes in physics, chemistry, and biology, ranging from X-ray astrophysics to DNA repair. Understanding and controlling these mechanisms are long-standing goals. For experiments on charge rearrangement it is crucial to locate the initial excitation and to trace the change in electron density over time. The advent of XUV and X-ray FELs has opened up exactly this possibility: The strong spatial localization of XUV radiation and X-rays combined with femtosecond pulse durations enables ultra-fast pump-probe experiments either with a delayed replica of the FEL pulse itself or a synchronized optical laser.

Here, we use an XUV-pump–XUV-probe scheme to access electron-rearrangement dynamics in dissociating molecular iodine ions. A preceding pulse of 87 eV, delivered by FLASH, multiply ionizes and consequently fragments iodine molecules (I_2). Depending on its temporal delay with respect to the pump pulse, an identical probe pulse may induce electron transfer between the dissociating ions. For small delays and correspondingly small internuclear distances symmetrically charged fragments are observed because an induced charge asymmetry is balanced by electron transfer. In contrast, electron transfer is blocked for large delays and large internuclear separations and thus we observe asymmetrically charged ion pairs.

By means of a Reaction Microscope we record the yield of the various coincident ion pairs from dissociating multiply charged molecular ions as a function of the pump-probe delay. This way, we are able to determine the critical time scales and internuclear distances up to which electron transfer is possible for the different occurring molecular break-up channels. Exemplarily, the results for coincident ion pairs $I^+ + I^{n+}$ with $n > 2$ are shown in Fig. 1 Our results are in very good agreement with predictions of a classical over-the-barrier model demonstrating its validity in the so far unexplored low-energetic energy regime relevant for FEL, plasma and chemistry applications.

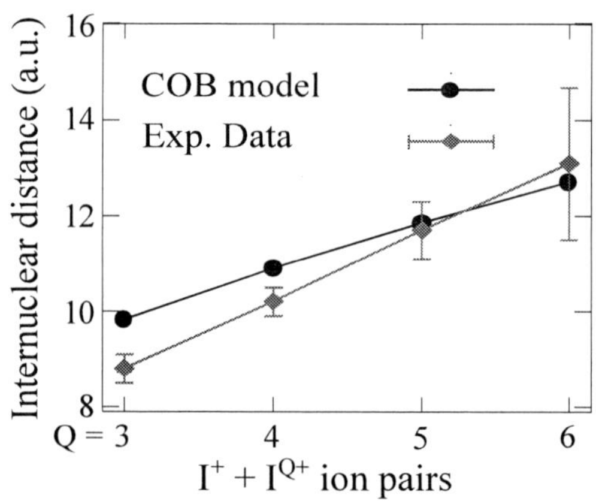

Figure 1. The critical internuclear distances over which electron transfer is possible are shown for all measured asymmetrically charged ion pairs with one ion being I^+. The experimental data are plotted as red triangles and the predictions of the classical over-the-barrier model as black dots.

[1]E-mail: kirsten.schnorr@mpi-hd.mpg.de

Exploring deep-core photoionization : ion-electron correlation effects in the 1 to 10 keV x-ray region

R. Guillemin*[1], L. Journel*, T. Marchenko*, T. Marin*, M.N. Piancastelli†, and M. Simon*

*Sorbonne Universités, UPMC and CNRS, UMR 7614, Laboratoire de Chimie Physique Matière et Rayonnement, F-75005, Paris, France

† Department of Physics and Astronomy, Uppsala University, PO Box 516, SE-751 20 Uppsala, Sweden

Synopsis We will report recent results on correlation effects in single-photon multiple-ionization processes after deep-core level excitation in the 1 to 10 keV photon energy range.

Understanding of the processes resulting from interaction between light and matter is of interest in many diverse fields such as photochemistry, astrophysics, as well as in biology and medicine. The lifetime of excited electronic states determines the time scale on which processes following the photoexcitation of a system, an atom, or a molecule will take place. Core-level photoexcitation brings this time scale down to the femtosecond, as core-hole lifetimes are typically a few femtoseconds for shallow core levels in the soft x-ray region and 1 fs or less for deep core levels, and thus provides an internal clock that allows the study of ultrafast phenomena. Studies of photoemission processes induced by hard x-rays have become feasible due to recent substantial improvement of instrumentation. Novel dynamical phenomena have become possible to investigate in this new regime. Moreover, photoionization of deep core shells is usually followed by a number of decay pathways, involving x-ray emission and multiple Auger decays which lead to the formation of multiply charged ionic states.

In this report, we will present recent results where correlation effects between electrons, ions and fragmentation dynamics are evidenced.

One possible way to look at the dynamics of photoemission is to study the effects of post-collision interaction (PCI) on the line shape of the photoelectron and the Auger electron peaks. Measurements of Auger electrons and ion/photoelectron coincidences after argon 1s photoionization provide new information on the dynamics of Auger relaxation on the femtosecond timescale [1].

In the photoionization process, the outgoing electron carries a momentum that is identical but opposite to the momentum of the remaining atom. High-energy photoemission at energies well away from any resonances can significantly affect Auger emission. In particular, we have observed a phenomenon showing a marked variation in Auger lineshape, depending on the direction with respect to the detector, which we interpret as a manifestation of the Doppler effect due to ion recoil after emission of an energetic photoelectron [2].

Finally, using a joint experimental and theoretical study of core-ionized carbon disulfide (CS_2), we demonstrate that it is possible to experimentally select distinct molecular-fragmentation pathways in which the core hole can be considered as either localized on one sulfur atom or delocalized between two indistinguishable sulfur atoms. This feat is accomplished by measuring photoelectron angular distributions within the frame of the molecule [3].

References

[1] R. Guillemin *et al* 2012 *Phys. Rev. Lett.* **109** 013001

[2] M. Simon *et al* 2014 Nature Communications **5** 4069

[3] R. Guillemin *et al* 2015 Nature Communications **6** 6166

[1] E-mail: renaud.guillemin@upmc.fr

ph-I7

Photoionization induced fragmentation of glycine molecule and endohedral fullerenes $Ho_3N@C_{80}$ molecule

L. Fang [1, a], H. Xiong [2], T. Osipov [3], V. S. Petrovic [4], T. Wolf [4], E. Sistrunk [4], M. Gühr [4] and N. Berrah [2]

[1]Center for High Energy Density Science, University of Texas, Austin, TX 78712, USA
[2]University of Connecticut, Physics Department, Storrs, CT 06269, USA
[3]LCLS, SLAC, Menlo Park, CA 94025, USA
[4]PULSE Institute, SLAC, Menlo Park, CA 94025, USA

Synopsis We present our investigation on the photoionization and the fragmentation of glycine molecule and $Ho_3N@C_{80}$ molecule using x-ray light source and femtosecond optical laser respectively. Following the decay of core-hole states of glycine, the molecule fragments via various pathways which are identified by ion coincidence measurements. We compare the branching ratios of these different fragmentation channels which are shown to be dependent on the photon energy. We found that the ionization of $Ho_3N@C_{80}$ molecule with ultrafast optical lasers has a different mechanism than that of C_{60} molecule. At high laser intensity, a rapid nuclear frame heating mechanism is also uncovered.

Molecular dissociation following photoionization provides insights into the photo-interaction dynamics as well as the ionization mechanisms [1, 2]. We investigated photoionization and fragmentation of glycine molecule and $Ho_3N@C_{80}$ molecule induced with x-ray light source and femtosecond optical laser respectively.

The experiment on glycine was carried at the Advanced Light Source (ALS) at the Lawrence Berkley National Laboratory (LBNL). The photon energies were judiciously chosen to core - ionize the carbon and the oxygen atoms in the glycine molecule. Ion coincidence measurements show various dissociation channels and different dissociation patterns associated with different atomic sites of the photoionization. In the experiment on $Ho_3N@C_{80}$ molecule, laser pulses of 30 fs duration from a Ti:Sapphire laser system was used to photoionize the molecule. Power law of the ionization was investigated and compared to that of C_{60} molecule [3, 4]. The comparison indicates the different ionization mechanism due to the difference in energy levels and molecular structures. In both experiments, ion signals were recorded by a velocity map imaging (VMI) spectrometer with a hexanode delay-line detector which enables there-dimensional momentum measurement.

We show that the branching ratios of various fragmentation channels of glycine molecule vary as the x-ray photon energy changes, i.e., for different core-hole locations and final ionic states. The ion yield difference pattern between different photon energies implies bond breaking trends. Bonds that preferentially break up are not necessarily at where the initial core hole is located. The fragmentation of $Ho_3N@C_{80}$ shows bond breaking within the carbon cage and the core molecule inside the cage, and also formation of bonds between the cage fragment and the core fragments. The kinetic energy distribution of the Ho^+ ions were obtain and shows a rapid ionization mechanism of the carbon cage with laser pulses of high intensities.

This work was funded with DOE-BES, No. DE-FG02-92ER14299.

References

[1] L. Fang *et al* 2012 *Phys. Rev. Lett.* **109** 263001
[2] L. Fang *et al* 2014 *J. Phys. B: At. Mol. Opt. Phys.* **47** 124006
[3] B. F. Murphy *et al* 2014 *Nat. Commun.* **5** 4281
[4] N. Berra *et al* 2015 *Journal of Modern Optics.* Submitted.

[a] E-mail: lifang@austin.utexas.edu

Electronic and Nuclear Dynamics Triggered by Ultra-Intense Soft and Hard X-rays

Artem Rudenko* [1]

*J.R. Macdonald Laboratory, Department of Physics, Kansas State University, Manhattan KS, USA

Synopsis We present the results of recent experiments on multiphoton ionization of atoms and molecules by ultra-intense femtosecond X-ray pulses at the photon energies ranging from 1.5 to 8.3 keV. We observe unprecedentedly high charge states of heavy atoms (up to 48+ in Xe at 8.3 keV), found signatures of efficient charge redistribution in molecules, and trace its mechanisms and spatial range in a dedicated laser pump – X-ray probe experiment.

The development of high-intensity, short-pulsed X-ray radiation sources promises revolutionary new techniques in diverse scientific fields, nurturing the vision of dynamic imaging of matter with angstrom spatial and femtosecond temporal resolution [1]. In particular, the start-up of the first hard X-ray free-electron laser, the Linac Coherent Light Source (LCLS), triggered a variety of imaging experiments on small molecules, clusters, nanocrystals, and isolated nano-scale particles. The basic prerequisite for designing these experiments is understanding the response of individual atoms, and tracing electronic and nuclear dynamics in the vicinity of the atom that absorbed X-ray photon(s), thus, revealing basic mechanisms of local radiation damage. Pursuing this goal, we report on a series of experiments on multiphoton multiple X-ray ionization of isolated heavy atoms and similar atoms embedded into molecular systems.

For soft X-ray ionization of heavy rare gas atoms such as Xe [2] and Kr [3], we observed the charge states well above the predictions of the simple sequential ionization model (up to 36+ instead of 26+ for Xe at 1500 eV), which was very successful in explaining results of the earlier LCLS experiments on lighter elements [4]. Comparing experimental data with theory (XATOM model [5]), studying the wavelength dependence and the fluorescence spectra, we found that intermediate resonant excitations are responsible for the extremely high charge states observed. Comparing the ionization of the isolated Kr and Xe atoms with the ionization of molecules containing single Se (CH_3SeH, C_2H_5SeH) or I (ICl, CH_3I) atom, which have the electronic structure very similar to Kr and Xe, respectively, we found signatures of efficient charge redistribution interweaved with ultrafast nuclear dynamics [6,7]. In a dedicated laser pump / X-ray probe experiment, we studied similar charge rearrangement reaction in a laser-dissociated CH_3I molecule, showed that its spatial range can be estimated from a classical over-the-barrier model and observed signatures of electron transfer up to internuclear separations of ~ 20 Å [8].

We complement the results in the soft X-ray domain with the data obtained using ultra-intense hard (5.5-8.3 keV) X-ray radiation. By focusing the LCLS beam to ~ 0.1 μm^2 spot, we achieve the unprecedented X-ray intensities of ~ 10^{20} W/cm^2. Under these conditions we were able to strip all 18 electrons from argon atoms, 34 (all but two 1s) electrons from krypton, and reached the record 48+ charge state for xenon at 8.3 keV. Although the highest charge states is reached at the highest photon energy (8.3 keV), the wavelength dependence of the spectra from 5.5 to 8.3 keV shows significant overall enhancement in the production of high charge states at the intermediate photon energies (around 6.5 keV). Similar to the soft X-ray regime, this behaviour results from the ionization enhanced or even enabled by intermediate resonance excitations [2,3].

For small molecules, in contrast to our results in the soft X-ray regime [6-8], the presence of molecular partners does not significantly reduce the highest charge state observed for high-Z elements: we detected I^{47+} ions from CH_3I at 8.3 keV as compared to Xe^{48+}. However, for larger systems like C_6H_5I, the maximum charge state observed is ~ I^{30+}, indicating that the charge transfer from the neighbouring atoms still plays an important role if enough electrons are present. In both cases, the highest charge state of carbon observed (C^{4+}) is in good agreement with the over-the-barrier charge transfer model used in [8].

References

[1] J. Ullrich, A. Rudenko and R. Moshammer 2012 *Annu. Rev. Phys. Chem.* **63** 635
[2] B. Rudek *et al* 2012 *Nature Photonics* **6** 858
[3] B. Rudek *et al* 2013 *Phys. Rev. A* **87** 023413
[4] L. Young *et al* 2010 *Nature* **466** 56 (2010).
[5] S.K. Son and R. Santra 2012 *Phys. Rev. A* **85** 063415
[6] B. Erk *et al* 2013 *Phys. Rev. Lett.* **110** 053003
[7] B. Erk *et al* 2013 *J. Phys. B* **46** 164031
[8] B. Erk *et al* 2014 *Science* **345** 288

[1] E-mail: rudenko@phys.ksu.edu

ph-I9

Photoelectron diffraction from laser-aligned molecules using an x-ray free-electron laser

Kyo Nakajima* [1], Takahiro Teramoto†, Hiroshi Akagi‡, Takashi Fujikawa§, Takuya Majima|, Shinichirou Minemoto¶, Kanade Ogawa**, Hirofumi Sakai¶, Tadashi Togashi††, Kensuke Tono††, Shota Tsuru§, Ken Wada*, Makina Yabashi**, and Akira Yagishita* [2]

* Institute of Materials Structure Science, KEK, 1-1 Oho, Tsukuba 305-0801, Japan
† College of Science and Engineering, Ritsumeikan Univ., 1-1-1 Noji-higashi, Kusatsu 525-8577, Japan
‡ Quantum Beam Science Center, JAEA, 8-1-7 Umemidai, Kizugawa 619-0215, Japan
§ Graduate School of Science, Chiba Univ., 1-33 Yayoi-cho, Inage, Chiba 263-8522, Japan
| Graduate School of Engineering, Kyoto Univ., Kyoto daigaku-katsura, Nishikyo, Kyoto 615-8530, Japan
¶ Graduate School of Science, The University of Tokyo, 7-3-1 Hongo, Bunkyo, Tokyo 113-0033, Japan
** RIKEN SPring-8 Center, 1-1-1 Kouto, Sayo-cho, Sayo-gun, Hyogo 679-5148, Japan
†† Japan Synchrotron Radiation Research Institute, 1-1-1 Kouto, Sayo-cho, Sayo-gun, Hyogo 679-5198, Japan

Synopsis We measured deep inner-shell *2p* X-ray photoelectron diffraction (XPD) patterns from laser-aligned I_2 molecules using X-ray free-electron laser (XFEL) pulses. The XPD patterns of the I_2 molecules, aligned parallel to the polarization vector of the XFEL, were well matched with our theoretical calculations. We have demonstrated that this approach is a significant step toward the time-resolved imaging of molecular structures.

X-ray photoelectron diffraction (XPD) with the intense and ultrafast pulses from X-ray free-electron lasers (XFELs) is a promising method for the determination of the geometric structure of molecules on an ultrafast time scale in pump-probe arrangements [1]. Toward the achievement of this goal, we have investigated I *2p* XPD from laser-aligned I_2 molecules with XFEL pulses.

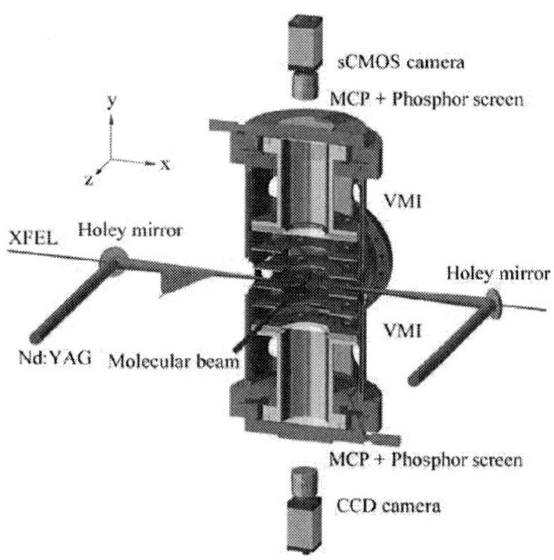

Figure 1. Schematic drawing of the experimental set-up. Two laser beams propagating in a collinear arrangement intersect a supersonic pulsed molecular beam at the center of a vacuum chamber.

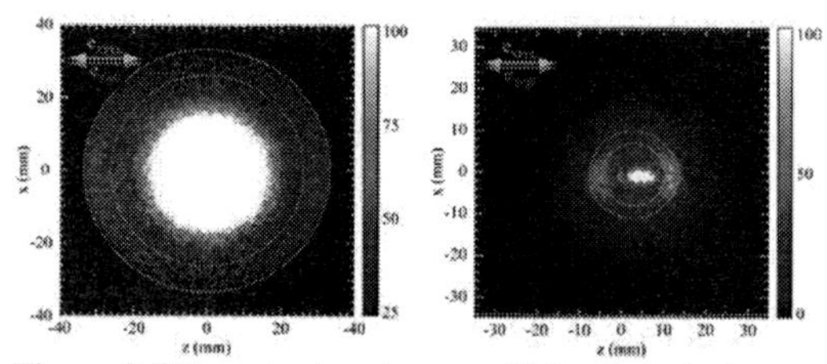

Figure 2. 2D momentum images of I *2p* photoelectrons with an energy of 140 eV (upper panel) and fragment ions (lower panel) from the laser-aligned I_2 molecules. The directions of polarization of YAG laser and XFEL are indicated by double-headed arrow.

The experiment has been performed using our apparatus consists of velocity-map imaging spectrometers (Fig. 1) at the SPring-8 Angstrom Compact Free-Electron Laser (SACLA) [2]. We have successfully measured 2D momentum images of I *2p* photoelectrons and fragment ion pairs from the I_2 molecules, aligned parallel to the polarization vector of the XFEL [3]. The XPD patterns were in strong agreement with the multiple-scattering XPD theory calculations. This is a step toward ultra-fast photoelectron diffraction to determine molecular structures. In this talk, the details of the experimental and theoretical results will be discussed.

References

[1] M. Kazama *et al* 2013 *Phys. Rev. A* **87** 063417
[2] T. Ishikawa *et al* 2012 *Nat. Photon* **6** 540
[3] K. Nakajima *et al Scientific Reports* submitted

[1] Present address: JASRI, Sayo-cho, Sayo-gun, Hyogo 679-5198, Japan. E-mail: kyo.nakajima@spring8.or.jp
[2] E-mail: akira.yagishita@kek.jp

Attosecond Spectroscopy: from measuring ionization times to time-resolving chiral response

Olga Smirnova [1]

[1]Max Born Max-Born-Institute, Max-Born-Str. 2A, 12489 Berlin , Germany

Ionization of atoms and molecules in strong infrared laser fields is a fundamental process that plays key role in many research fields, from attosecond physics to laser modification of transparent materials. I will describe new results related to this process.

I will first introduce protocols that can be used to define and measure time delay associated with strong-field ionization, and will discuss how this time depends on the number of photons required to overcome the ionization potential. Since strong field ionization is often viewed as electron tunnelling through the barrier created by the superposition of the atomic binding potential and the laser field, this question touches upon the famous problem of time delays in tunnelling. Using the combination of analytical theory and ab-initio numerical simulations, I will show that optical tunnelling is instantaneous [1], at least within the framework of non-relativistic quantum mechanics.

I will then show that optical tunnelling in polyatomic molecules populates different hole states, inducing correlated electron-hole dynamics. Importantly, in chiral molecules the hole dynamics becomes chiral, sensitive to the helicity of the driving laser field and the chirality of the molecule. The infrared laser field that removes the electron from the molecule can also bring this electron back to the parent ion, where it can recombine with the hole left behind. Light emitted during recombination takes the snapshot of the recombining system, recording the electron-hole dynamics with attosecond temporal resolution. I will show that in chiral systems the emitted light distinguishes left-handed and right-handed molecules with extreme sensitivity and records the chiral dynamics of the hole with about 0.1 fsec temporal resolution [2].

References

[1] L. Torlina, F. Morales, J. Kaushal, I. Ivanov, A. Kheifets, A. Zielinski, A. Scrinzi, H. G. Muller, S. Sukiasyan, M. Ivanov and O. Smirnova, Interpreting attoclock measurements of tunnelling times, *Nature physics*, 2015

[2] R. Cireasa, A. E. Boguslavskiy, B. Pons, M. C. H. Wong, D. Descamps, S. Petit, H. Ruf, N. Thire, A. Ferre, J. Suarez, J. Higuet, B. E. Schmidt, A. F. Alharbi, F. Legare, V. Blanchet, B. Fabre, S. Patchkovskii, O. Smirnova, Y. Mairesse, V. R. Bhardwaj, Probing molecular chirality on sub-femtosecond time-scale, *Nature physics*, 2015

ph-I11

Circular dichroism in two-color multi-photon ionization of rare-gas atoms

T. Mazza[1*], M. Ilchen[1], A.J. Rafipoor[1], C. Callegari[2], P. Finetti[2], O. Plekan[2], K.C. Prince[2], R. Richter[2], M.B. Danailov[2], A. Demidovich[2], G. De Ninno[2], C. Grazioli[2], R. Ivanov[2], N. Mahne[2], L. Raimondi[2], C. Svetina[2], L. Avaldi[3], P. Bolognesi[3], M. Coreno[3], P. O'Keeffe[3], M. Di Fraia[4], M. Devetta[5], Y. Ovcharenko[6], Th. Möller[6], V. Lyamayev[7], F. Stienkemeier[7], S. Düsterer[8], K. Ueda[9], J.T. Costello[10], E V Gryzlova[11], A N Grum-Grzhimailo[11], A.K. Kazansky[12], N.M. Kabachnik[1,12], M. Meyer[1]

[1]European XFEL GmbH, Albert-Einstein-Ring 19, D-22761 Hamburg, Germany; [2]Elettra-Sincrotrone Trieste, Basovizza I-34149, Italy; [3]CNR Istituto di Metodologie Inorganiche e dei Plasmi, CP10, Monterotondo Scalo I-00016, Italy; [4]Department of Physics, University of Trieste, Trieste I-34128, Italy; [5]CIMAINA and Dipartimento di Fisica, Università di Milano, via Celoria 16, Milano I-20133, Italy; [6]Technische Universität Berlin, Hardenbergstrasse 36, D-10623 Berlin, Germany; [7]Physikalisches Institut, Universität Freiburg, D-79104 Freiburg, Germany; [8]Deutsches Elektronen-Synchrotron (DESY), Notkestrasse 85, D-22607 Hamburg, Germany; [9]Institute of Multidisciplinary Research for Advanced Materials, Tohoku University, Sendai 980-8577, Japan; [10]National Centre for Plasma Science and Technology, Dublin City University, Dublin 9, Ireland [11]Skobeltsyn Institute of Nuclear Physics, Lomonosov Moscow State University, Moscow 119991, Russia [12]Donostia International Physics Center (DIPC), San Sebastian/Donostia E-20018, Spain

Synopsis We report on the study of two-color multiphoton ionization using circularly polarized XUV and IR radiation. The angle-resolved photoelectron spectra allow investigating the chirality of continuum and resonantly excited electronic states. The same two-color ionization scheme was used to characterize the polarization state of the radiation delivered by the FERMI FEL.

2-color experimental techniques involving ultra-short XUV and optical light are powerful tools for understanding the dynamics of electronic quantum systems and in addition have been demonstrated to be effective for the characterization of intense femtosecond x-ray radiation.

The availability of circularly polarized FEL radiation allows the tuning of an additional key parameter in this class of experiments, opening to the investigation of ultrafast dynamics in chiral samples such as magnetic systems or functional complexes with biomedical applications. In addition, it calls for an effective metrology tool for the polarization state of the radiation.

In this report we show how a straightforward 2-color ionization scheme of helium atoms allows an accurate measurement, independent on photon energy, of the circular polarization degree of a free electron laser beam using circular dichroism [1]. This approach allowed us to assess that radiation with a high (>90%) degree of circular polarization is delivered to the experiment at FERMI, consistent with pure source properties and with the characteristics of the beam transport system.

Two-color multiphoton ionization with circularly polarized light was investigated focusing on the photoelectron angular distribution under different polarization conditions [2]. The comparison of the experimental results with different theoretical descriptions based on Strong Field Approximation and time-dependent perturbation theory allows to access information on the contribution of different electronic partial waves to the ionization process.

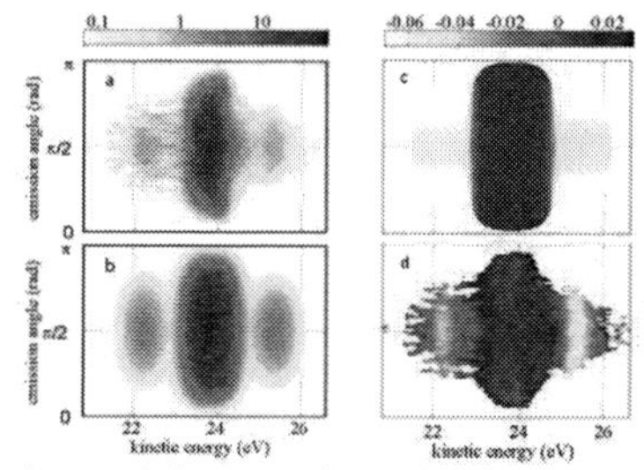

Experimental (a,c) vs theoretical (b,d) cross section (a,b) and circular dichroism (c,d) in He 1s 2color photoionization. The dichroism is a direct measurement of the degree of circular polarization.

Thanks to the spectral properties of the FERMI seeded FEL, it was also possible to study the chirality of resonantly excited states in ionic atomic species created by circularly polarized XUV light and probed by optical radiation [3]. The first results of these investigations will be presented.

References

[1] T. Mazza et al., *Nature Communications* **5** 3648, (2014). [2] T. Mazza et al., *New Journal of Physics*, under review (2015). [3] M. Ilchen et al., *manuscript under preparation*.

*E-mail: tommaso.mazza@xfel.eu

Dichroism in two-color ionization of atoms by short pulses. ph-I12

Kazansky A K [1]

IKERBASQUE, Basque Foundation for Science, E-48011 Bilbao, Spain;
Departamento de Fisica de Materiales, UPV/EHU, E-20018 San Sebastian/Donostia, Spain;
Donostia International Physics Center (DIPC), E-20018 San Sebastian/Donostia, Spain

Synopsis Quite recently, circularly polarized short pulses (in few tens of fs duration range and in XUV and X-ray energy range) have been obtained at FERMI [1] and LCLS FEL sources. It makes possible to investigate a set of various effects related to circular dichroism. The simplest phenomenon to be studied is dichroism in two-color ionization of atoms. The corresponding theory will be presented in the talk. It is based on the Strong Field Approximation and is valid if the energy of the ionizing XUV pulse, is well above the ionization threshold. The theory allows one to obtain a full description of all the features of the ionization process in this energy region.

The ionization of unpolarized atom by collinear circularly polarized XUV and IR pulses is theoretically considered. The pulses supposed to be either co-polarized, i.e. the electric fields are rotating in the same direction, or counter-polarized. The goal of the study is obtaining double differential cross sections (DDCS) for both possible relative polarizations of the beams, $\sigma_{co,count}(\vec{k}_f)$. The relative quantity, circular dichroism is defined as $CD = \left(\sigma_{co}(\vec{k}_f) - \sigma_{count}(\vec{k}_f)\right) / \left(\sigma_{co}(\vec{k}_f) + \sigma_{count}(\vec{k}_f)\right)$.

In the talk, the XUV pulse will be considered assuming its frequency to be large enough, so that the initial energy of the ionized electron $E_{init} = \omega_X - |E_b|$ just after absorption of XUV photon is above 20 eV. Duration of the XUV pulse is assumed to be in a few tens fs range. The IR laser pulse has the conventional frequency 1.55 eV (800 nm) and, as a rule, is longer than XUV pulse. In most of the cases, the relative delay of the pulses is not relevant.

The basic quantity of the theory, the ionization amplitude, within the Strong Field Approximation (SFA) reads [2], [3]:

$$\mathcal{A}(\vec{k}_f) = \int_{-\infty}^{\infty} dt \mathcal{E}_X(t) Y_\alpha(\vec{k}_0(t)) \mathcal{D}(k_0(t)) \exp(i\Phi(t)),$$

where $\vec{k}_0(t) = \vec{k}_f - \vec{A}_{IR}(t)$ is the electron momentum just after ionization, $\vec{A}_{IR}(t) = A_0(t)(\vec{n}_x \cos(\omega t) \pm \vec{n}_y \sin(\omega t))$ is the vector-potential of circularly polarized IR field, $\mathcal{E}_X(t)$ and $A_0(t)$ are the corresponding pulses envelops, and $\Phi(t) = \int_t^\infty dt' \left([\vec{k}_0(t)]^2 - \omega_X + |E_b|\right)/2$ is the Wolkov phase, $\mathcal{D}$ is the dipole matrix element.

The SFA provides a very simple, versatile, and convenient computational scheme, which allows universally to compute the amplitudes of all the processes in the two-color experiments. However, it is noteworthy that SFA implies that the electron leaves the region where Coulomb field is important quite promptly. This approximation is not valid when the energy E_{init} is close to the ionization threshold.

The spectrum of the ejected electrons consists of a set of relatively narrow peaks (sidebands) corresponding to absorption of the XUV photon and of an integer number of IR photons. The results of computations of DDCS and CD for ionization of He atom, when the initial state is $1s$ state [2], [3], and for Ne [4], when the initial state is $2p$ state, will be reported. Also the results obtained in analysis of the first two-color experiment [1] on FEL with circularly polarized light will be presented.

Besides the circular polarization, the SFA approach allows one to study the so-called linear dichroism, when both pulses are linearly polarized and two cases are considered with two perpendicular orientations of polarization of one of the pulses. This case will be addressed in the talk [4]. Also, analytical approximation which is valid for quite long IR pulse, will be presented. These formulas quite well reproduce the behaviour of DDCS in comparison with the results of numerical calculations.

References

[1] T.Mazza *et al* 2014 *Nat. Comm.* **5** 4648

[2] A.K.Kazansky *et al* 2011 *Phys. Rev. Lett* **107** 253002

[3] A.K.Kazansky *et al* 2012 *Phys. Rev. A* **85** 053409

[4] A.K.Kazansky *et al* 2014 *J. Phys.B* **47** 065602

[1]E-mail: kazan356@rambler.ru

ph-I13

Determination of spin-polarized H atoms produced in molecular photodissociation

Bernadette M. Broderick[#], Oleg S. Vasyutinskii*[1], Arthur G. Suits[#]

[*] *Ioffe Institute, St.Petersburg, 194021 Russia*
[#] *Department of Chemistry, Wayne State University, Detroit MI 48202 USA*

Sinopsis We demonstrate that a two-color high-n Rydberg excitation of hydrogen atoms using polarized tagging laser may be combined with pulsed-field ionization to allow for direct measurement of the velocity dependence of hydrogen atom spin polarization with high resolution and sensitivity. The method can be applied for the study of dissociation of polyatomic molecules and reactive and inelastic scattering and promises a wealth of new detail concerning these elementary processes.

The production of spin-polarized hydrogen atoms in molecular photodissociation was considered theoretically more than thirty years ago [1], however it has been experimentally observed only recently, first by inferring based on the measured co-fragment angular momentum polarization [2], and subsequently by detection of the polarized fluorescence in the H atoms excited by Lyman-alpha radiation [3].

We present a new experimental technique allowing for direct measurement of the velocity dependence of hydrogen atom spin polarization with high resolution and sensitivity. The strategy is an adaptation of the H atom Rydberg time-of-flight approach (SP-HRTOF), where the spin-sensitive probe scheme combines a linearly polarized Lyman-α laser, a circularly polarized tagging laser, and a photolysis laser [4].

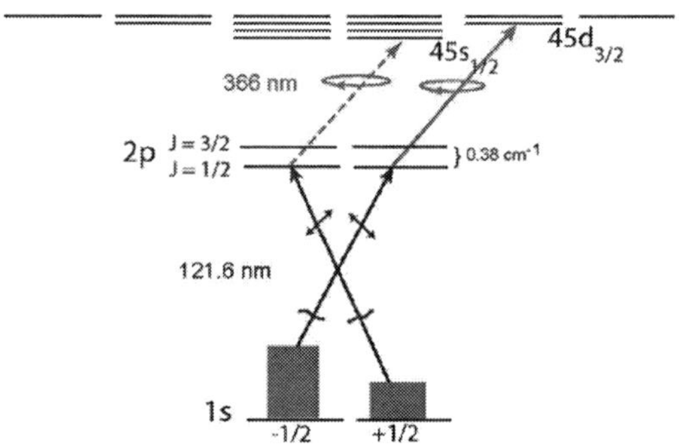

Figure 1. Spin selective Excitation Scheme

The approach is illustrated with a measurement of the H atom spin polarization in photodissociation of HBr and DBr at 212.8 nm. The two coherent contributions to the spin polarization are measured for H atoms produced in conjunction with dissociation to $Br(^2P_{3/2})$ and $Br^*(2P_{1/2})$; they are found to be negligible for the former channel but substantial for the latter, in agreement with previous theoretical predictions. The ratio of the two measurements directly gives the asymptotic scattering phase shift for dissociation along the two potentials for the H + Br* channel.

The paper discusses the adaptation of the method to detect the incoherent contribution to the photofragment spin polarization by using excitation of hydrogen atoms with circularly polarized tagging laser combined with pulsed-field ionization which allows for direct velocity map imaging of the H atom spin polarization. We also discuss the progress in applying the technique to dissociation of polyatomic molecules.

The presented results show the capability for SP-HRTOF to yield fundamental insights into molecular scattering processes. Applications to dissociation of polyatomic molecules and reactive and inelastic scattering promise a wealth of new detail concerning these elementary processes.

References

[1] O. S. Vasyutinskii 1981 *Sov. Phys. JETP* **54**, 855.
[2] T. P. Rakitzis *et al* 2003 *Science* **300**, 1936.
[3] D. Sofikitis *et al* 2008 *The Journal of Chemical Physics* **129**, 144302.
[4] B. M. Broderick *et al* 2013 *The Journal of Physical Chemistry Letters* **4**, 3489.
[5] B. M. Broderick *et al* 2014 *Review of Scientific Instruments* **85**, 053103.

[1] E-mail: osv@pms.ioffe.ru

ph-I14

Molecular dynamics on laser-controlled transition states

Andreas Fischer* [1], Martin Gärttner*, Philipp Cörlin*, Alexander Sperl*, Michael Schönwald*, Tomoya Mizuno*, Giuseppe Sansone†, Arne Senftleben‡, Joachim Ullrich§, Bernold Feuerstein*, Thomas Pfeifer*, Robert Moshammer* [2]

* Max-Planck-Institut für Kernphysik, Saupfercheckweg 1, 69117 Heidelberg, Germany
† Dipartimento di Fisica, Politecnico Milano, Piazza Leonardo da Vinci 32, 20133 Milano, Italy
‡ Institut für Physik, Universität Kassel, Heinrich-Plett-Str. 40, 34132 Kassel, Germany
§ Physikalisch-Technische Bundesanstalt, Bundesallee 100, 38116 Braunschweig, Germany

Synopsis Molecular wave-packet dynamics on laser-controlled transition states has been studied in a kinematically complete experiment. Emphasis is put on the behavior of a well-defined wave-packet in H_2^+, consisting of exactly two vibrational states, when it approaches the laser-dressed potential barrier between bound and dissociating states. Controlling the relative phase (delay) between the wave packet and the laser barrier and by comparing the phase dependent energies of fragments with quantum as well as semi-classical calculations, we identify an intuitive "elevator" effect that may be used to control the dynamics in chemical reactions.

Laser-induced dissociation is an important tool for manipulating photochemical reactions of molecular systems [1, 2]. Here, nuclear wave-packets are launched in H_2 by ionization with an attosecond pulse train and the molecule is subsequently dissociated by a time-delayed infrared-probe pulse. With this scheme, we obtain time-delay dependent momentum spectra of the fragments, showing that the reaction kinematics can be controlled by variation of the time-delay (see Fig. 1).

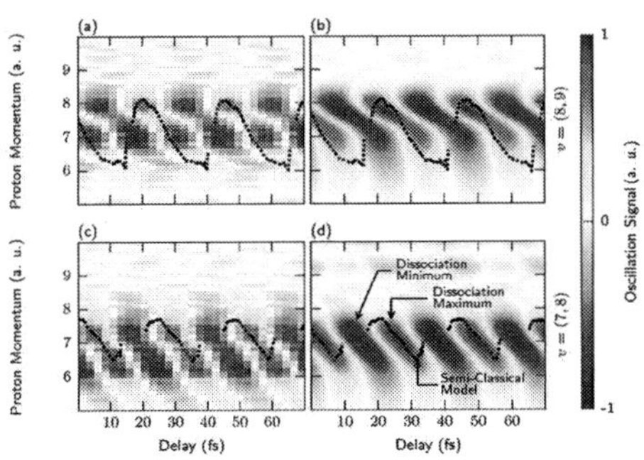

Figure 1. Wavelet analysis of the wave-packet dynamics, where for each picture only two vibrational states contribute. Left column: experimental data, right column: TDSE simulation. The dotted lines are the predictions of the semi-classical model.

Using a semi-classical model, supported by a quantum calculation, we are able to explain the time-dependency of the reaction kinematics in terms of an intuitive "elevator" effect [3], as the fragments are lifted or lowered (gain or loss of final momentum) on the laser-controlled potential energy surfaces (see Fig. 2).

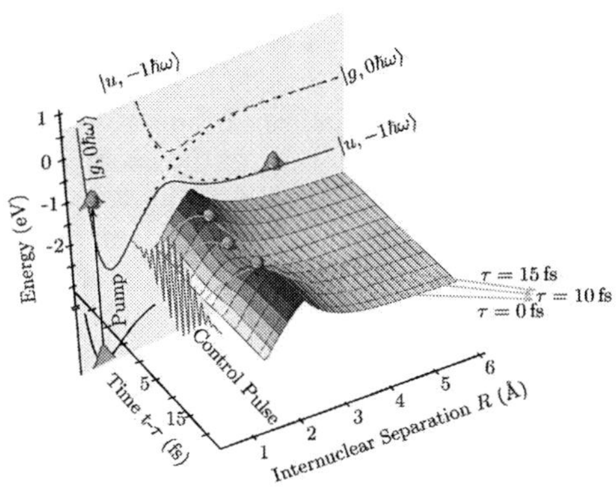

Figure 2. Time-delay dependent, semi-classical dissociation trajectories for three distinct cases: (i) gain of energy, (ii) constant energy and (iii) energy loss.

This semi-classical model is capable of describing and predicting the time-dependency of the momentum yield of the dissociation connected to a wave packet consisting of only two vibrational states. The fact that the wave-packet dynamics for light molecules such as H_2^+, where quantum effects are expected to be the most significant, can be understood in terms of semi-classical modeling indicates a broad applicability of such processes in quantum control [1, 2].

References

[1] Shapiro, M. *et al.*, 1997 *J. Chem. Soc. Faraday Trans.* **93, 12631277**

[2] Suessman, B. J. 2006 *Science* **314, 278-281**

[3] Fischer, A. *et al.* 2014 *arXiv*:**1410.8032**

[1] E-mail: andreas.fischer@mpi-hd.mpg.de
[2] E-mail: robert.moshammer@mpi-hd.mpg.de

ph-I15

Fragmentation of halopyrimidines and halouraciles by photoionization and ion impact

P. Bolognesi[1], M.C. Castrovilli[1], A. Kettunen[2], A. Cartoni[3], R. Richter[4], S. Tosic[5], S. Maclot[6], P. Rousseau[6], R. Delaunay[6], A. Domaracka[6], B. Huber[6] and L. Avaldi[1]

[1]CNR-Istituto di Struttura della Materia, Area della Ricerca di Roma1, Monterotondo Scalo, Italy
[2] Department of Physics, University of Oulu, Finland
[3] Dipartimento di Chimica, Sapienza Università di Roma, Roma, Italy
[4] Elettra-Sincrotrone Trieste, Basovizza, Italy
[5] Institute of Physics, University of Belgrade, Belgrade, Serbia
[6] CIMAP, CEA/CNRS/ENSICAEN, Université de Caen Basse-Normandie, Caen, France

Synopsis The fragmentation of halopyrimidines and halouraciles is studied by PEPICO experiments and ion induced mass spectrometry. The selectivity of the PEPICO experiments allows to define the role of the final cationic states, while the ion induced fragmentation of hydrated clusters sheds light on the tautomerisation that can lead to mutagenesis.

In recent years we have focused our interest in processes induced in isolated molecules by soft X-ray and ion beams. By photon ionization the site and state of the energy deposition are well defined and the following chain of processes are characterized by coincidence techniques. In the fragmentation induced by low energy ions, the effects of the increasing complexity of the target and the role played by the 'environment' on the properties of the single molecule are addressed.

In the presentation two examples will be discussed. The first one is the photofragmentation of 2Cl-pyrimidine following inner shell excitation or direct valence ionization studied via electron-ion coincidence techniques. The experiments have been performed at the Gasphase photoemission beamline at Elettra, Trieste (Italy). The results show that the resonant Auger process following inner shell excitation selectively populates the final states of the singly charged ion and the site- and state-selected fragmentation patterns appear to depend only on the final state of the singly charged ion. The comparison with state-selected photofragmentation after valence ionization confirms the role of the cationic state in the type of fragments produced.

In the second example the fragmentation of 5Br-uracil isolated molecules, homogeneous clusters and hydrated clusters by C^{4+} ions has been studied. The experiments have been performed at the ARIBE beamline of the GANIL facility, Caen (France). The observation of series of hydrated fragments provides the experimental evidence that a few water molecules attached to the 5Br-uracil can induce a tautomerisation process. This process can lead to mutagenesis and therefore to a different pairing in the DNA bases and can explain the radiosensitizing effect of compounds based on 5Br-uracil.

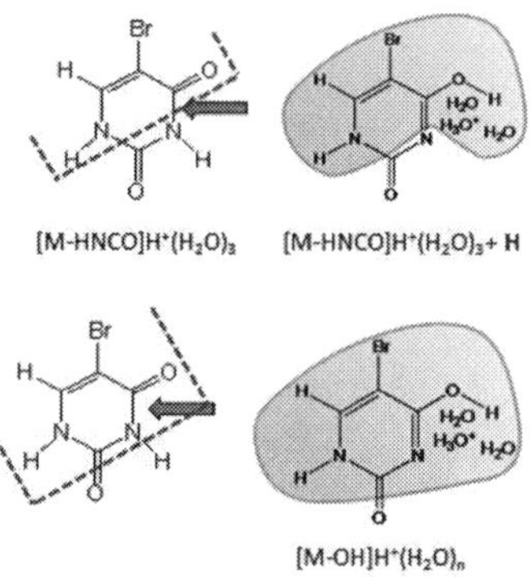

Figure 1: Schematic of the main fragments whose hydrated series are observed and assigned in the mass spectrum of hydrated clusters of 5Br-uracil. The blue arrows indicate the suggested site of hydration and the red dashed lines surround the charged (detected) fragment; M indicates the parent ion. The proposed tautomerisation processes mediated through the presence of a sufficient number of water molecules is also shown.

Acknowledgments

Work partially supported by the COST action XLIC, FIRB "Ion and neutral chemistry in the gas phase" and Italy-Serbia great relevance project "Nanoscale insights in radiation damage ".

[1]E-mail: Lorenzo.Avaldi@ism.cnr.it

Gas phase targets in interaction with Circularly-Polarized Light: Molecular polarimetry and chiroptical effects in photoionization

Laurent Nahon* [1]

*Synchrotron SOLEIL, Saint Aubin, France

Synopsis The interplay between molecular photoionization and elliptical/circular polarization is examined here in a dual context of polarimetry and of the study of chiral molecules.

Circularly Polarized Light (CPL) defines a 3D orientation of space with respect to which anisotropic properties of matter and of Matter/Light Interactions (MLI) can be studied. In particular, as a true chiral object, via the well-know chiral recognition process, CPL is sensitive to chirality giving rising to various types of Circular Dichroism (CD). Chirality can be intrinsic in the case of chiral molecules or extrinsic when the handedness is achieved by the combined system of photon and (fixed-in-space) molecule. The latter case is found in the so-called Circular Dichroism in the Angular Distribution (CDAD) as observed on Molecular Frame-Photoelectron Angular Distributions (MF-PADs) on achiral targets. Reciprocally matter may probe the polarization of light.

After a short introduction on chirality, CPL production by VUV synchrotron radiation and optical polarimetry measurement, we will explore two complementary facets of the matter/light polarization interplay:

1-known MLI chirality, unknown polarization: this will be illustrated by the determination of CDAD in MF-PADs derived from electron/ion Vectorial Correlation in the VUV dissociative ionization of NO, induced by a general elliptically polarized light [1]. From these measurements, it has been possible to determine the full set of Stokes parameter characterizing the polarization state of light, and especially to disentangle the polarized and unpolarized components of light. The accuracy of this "molecular polarimetry" method has been benchmarked versus optical polarization.

2- Know polarization, unknown MLI chirality: this will be illustrated by valence-shell Photoelectron Circular Dichroism (PECD), which is a CD in the angular distribution of photoelectrons produced by CPL-ionization of randomly-oriented pure enantiomers [2]. This effect is observed as a very intense (up to 35 %) forward/backward asymmetry with respect to the photon axis and which reveals the chirality of the molecule (configuration). PECD happens to be orbital-specific and photon energy dependent, and is a very subtle probe of the molecular potential, being very sensitive to static molecular structures such as conformers [3], chemical substitution [4], clustering [5] as well as to dynamical ones such as vibrational motion [6]. After an introduction to PECD, a few recent results will illustrate the properties of this chiroptical effect.

We believe that the study of these two facets is timely and interesting in a context of growing CD experiments performed with VUV CPL light from FELs [7] or HHG sources [8] and for which accurate polarimetry methods are needed [9].

References

[1] K. Veyrinas *et al.*, Phys. Rev. A **88**, 063411 (2013).
[2] L. Nahon *et al.*, J. Elec. Spec. Rel. Phen. (2015) DOI: 10.1016/j.elspec.2015.04.008.
[3] M. Tia *et al.*, J. Phys. Chem. A **118**, 2765 (2014).
[4] G. A. Garcia *et al.*, Phys Chem Chem Phys **16**, 16214 (2014).
[5] I. Powis *et al.*, Phys Chem Chem Phys **16**, 467 (2014).
[6] G. Garcia *et al.*, Nat. Commun. **4**, 2132 (2013).
[7] T. Mazza *et al.*, Nat. Commun. **5**, 3648 (2014).
[8] A. Ferré *et al.*, Nat. Photonics **9**, 93 (2015).
[9] E. Allaria *et al.*, Phys. Rev. X **4**, 041040 (2014).

[1] E-mail: laurent.nahon@synchrotron-soleil.fr

ph-I17

Momentum Exchange in Molecular Systems

Xiao-Jing Liu[*], Quan Miao[†,*,#], Minna Patanen[*], Oksana Travnikova[*,~], Grazieli Simões[*,□], Christophe Nicolas[*], Victor Kimberg[†,*], Yu-Ping Sun[†,°], T. Darrah Thomas[!], Witold Andrałojć[*], Hans Ågren[†], John D. Bozek[!!□], Kiyoshi Ueda[^], Faris Gel'mukhanov[*,†] and Catalin Miron[*,‡,1]

[*] Synchrotron SOLEIL, Saint-Aubin, France; [†]Theoretical Chemistry and Biology, School of Biotechnology, Royal Institute of Technology, Sweden; [‡]ELI-NP, " Horia Hulubei" National Institute for Physics and Nuclear Engineering, Măgurele, Jud. Ilfov, Romania; [#]College of Electronics, Communication and Physics, Shandong University of Science and Technology, Shandong, China; [°]School of Science, Shandong University of Technology, Shandong, China; [^]Institute of Multidisciplinary Research for Advanced Materials, Tohoku University, Sendai, Japan; [!]Department of Chemistry, Oregon State University, Oregon, United States; [□]Instituto de Quimica, Universidade Federal do Rio de Janeiro, Rio de Janeiro, Brazil; [~]Laboratoire de Chimie Physique-Matière et Rayonnement, Paris 06 University, Paris, France; [!!□]SLAC National Accelerator Laboratory, Menlo Park, USA.

Synopsis Three recent studies demonstrate that the momentum exchange in molecular systems carries accurate information about the ionization history.

Momentum exchange is a key notion in classical physics. The increasingly wider availability of third generation synchrotron facilities, such as SOLEIL synchrotron in France, together with always improving spectrometers and measurement methods, momentum exchange can be measured at the molecular level. For example, the rotational Doppler broadening (RDB) initially predicted theoretically by Sun et al. [1], was possible to experimentally confirm without ambiguity in molecular nitrogen [2].

Even if the RDB, similar to the translational Doppler broadening [3], broadens photoelectron spectra, there still remains a fundamental difference between them. These differences can be exploited to learn more about molecular orbitals anisotropy, when the molecular spectroscopy is carried out at an ultra-high resolution soft x-ray beamline, such as PLEIADES. Indeed, a fine analysis of the vibrationally-resolved X-ray photoelectron spectra of different states of N_2^+, recorded for different photon energies and orientations of the polarization vector, revealed clear dependencies of the spectral line widths on the X-ray polarization, as well as on the symmetry of the observed final electronic states [4].

Moreover, in X-ray electron spectroscopy, the photoelectron momentum is large, and the recoil angular momentum transferred to the molecule is larger when the photoelectron is ejected from a light atom compared to a heavy one. Using the textbook case of the hydrogen chloride molecule (HCl), for the first time a clear separation of overlapping rotational bands according to the ionization site was achieved, and thus allows one to monitor the molecular orbital (MO) extension at different atomic sites, and hence to map the composition of MOs [5].

A last example will demonstrate that a molecular system, like the oxygen molecule, can serve as workbench to the famous «Einstein-Bohr recoiling double-slit *gedanken* experiment», in which the momentum transfer between a particle and a recoiling slit could allow identification of which slit the particle passed through on its way to the detector, thus destroying the interference pattern. In our experiment, the neutral O_2 molecule is excited with soft x-ray synchrotron radiation into an unstable electronic state where the molecule dissociates into two atoms that are quickly (few fs) moving apart. Subsequent de-excitation of the system proceeds through the ejection of a fast (Auger) electron. Using a state-of-the-art electron-ion coincidence experiment [6], the momentum exchange between the emitted Auger electron and the molecular or the atomic ion, which plays the role of ultra-light, quantum slits, can be directly measured. For early decays, the momentum is transferred to the whole molecules (double-slit), but for a late Auger electron emission, the molecule is dissociated, and the emitted electron transfers the recoil momentum to only one of the two oxygen atoms. The asymmetric momentum transfer distinguishes the «path» (which slit the electron emerged from), and thus quenches the interference pattern [7].

References

[1] Y.-P. Sun *et al.* 2010 *Phys. Rev. A* **82** 052506
[2] T. D. Thomas *et al.* 2011 *Phys. Rev. Lett.* **106** 193009
[3] P. Baltzer *et al.* 1993 *Rev. Sci. Instrum.* **64** 2179
[4] Q. Miao *et al.* 2015 *J. Phys. Chem. Lett.* **6** 1568
[5] C. Miron *et al* 2014 *Nat. Comm.* **5** 3816
[6] C. Miron *et al* 2009 *Nucl. Instrum. Methods A* **601** 66
[7] X.-J. Liu *et al* 2015 *Nat. Photonics.* **9** 120

[1]E-mail: catalin.miron@eli-laser.eu

ORAL CONTRIBUTIONS

Low-energy (67 eV) electron-impact induced interatomic coulombic decay of argon dimer

Xueguang Ren[*‡1], **Elias Jabbour Al Maalouf**[†], **Stephan Denifl**[†], **Alexander Dorn**[‡]

[*] Physikalisch-Technische Bundesanstalt, 38116 Braunschweig, Germany
[‡] Max-Planck-Institute for Nuclear Physics, 69117 Heidelberg, Germany
[*] Institute for Ion and Applied Physics, University of Innsbruck, Technikerstrasse 25, 6020-Innsbruck, Austria

Synopsis We study the electron-impact induced ionizing fragmentation and interatomic processes of Ar dimer (Ar_2) at low projectile energy of 67 eV with the multiple-particle coincidence method using a reaction microscope. From the measured kinetic energies of final state electrons and ions, we identify two decay channels which can contribute to the double peak structure in the observed kinetic energy release (KER) spectrum of argon dimer dissociating into $Ar^+ - Ar^+$.

In the weakly bond systems, e.g. in the Ar dimer, due to the presence of the environment new autoionization channels as known as Interatomic or Intermolecular Coulombic Decay (ICD) can be opened by transferring the deposited excess energy to the neighbor ionizing it and resulting in highly reactive species: low-energy ICD electrons and a pair of energetic ions. Since its first prediction in 1997 by Cederbaum and co-workers [1], ICD soon led to a strong interest due to its important roles in various mechanisms of radiation biology and chemistry [2, 3]. An additional relaxation process named radiative charge transfer (RCT) may occur in the Ar dimer in which one atom is doubly ionized, and an electron is transferred from a neighbor to the doubly charged Ar^{2+} ion. The system can relax by emitting a photon. Both ICD and RCT processes can result in two repulsive Ar^+ ions.

It has been observed upon ion- [4], electron- [5], and photon- [6] collisions with Ar dimer that the KER spectrum of the $Ar^+ - Ar^+$ channel has a double peak structure. Calculations have shown that a contribution of this KER could be attributed to decay channel of ICD [7]. Other calculations suggest that RCT may also contribute in this KER [8].

In the present work, we study the ionization of Ar dimer by low-energy electron-impact using the multiple-particle coincidence technique [9] in which the kinetic energies of all final state particles are measured. The measured KER spectrum is presented in Figure 1, which shows two peaks: a peak at 3.8 eV and the other one at 5.2 eV. These two energies correspond respectively to the two internuclear distances of 3.8 angstrom and 2.8 angstrom in the Ar dimer [7, 8]. From the measured kinetic energies of final-state particles, the binding energy for each channel is obtained. Thereby, we can identify the two decay channels (ICD and RCT) in the Ar dimer. Detailed results will be presented at the conference.

[1]E-mail: xue.g.ren@ptb.de

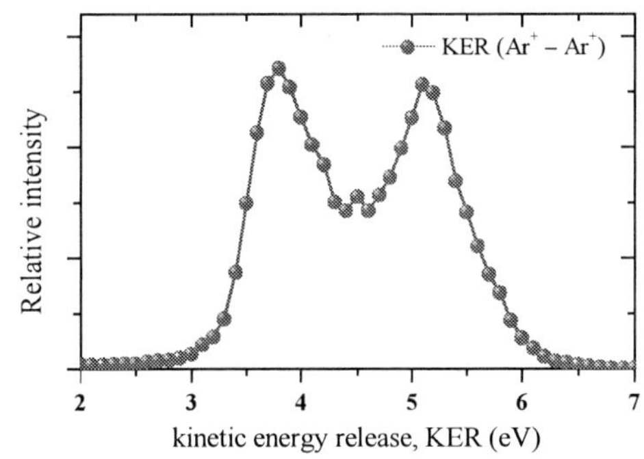

Figure 1. KER distribution for the $Ar^+ - Ar^+$ from the low-energy (67 eV) electron-impact ionization and dissociation of the $(Ar_2)^{2+}$.

References

[1] L. S. Cederbaum *at al* 1997 *Phys. Rev. Lett.* **79** 4778

[2] U. Hergenhahn 2011 *J. Elect Spec. Real. Phen.* **184** 7890

[3] T. Jahnke 2015 *J. Phys. B* **48** 082001

[4] H. Kim *at al* 2013 *Phys. Rev. A* **88** 042707

[5] S. Yan *at al* 2013 *Phys. Rev. A* **88** 042712

[6] M. Kimura *at al* 2013 *Phys. Rev. A* **88** 043414

[7] T. Miteva *at al* 2014 *J. Chem. Phys.* **141** 064307

[8] S. D. Stoychev *at al* 2008 *J. Chem. Phys.* **128** 014307

[9] X. Ren *at al* 2014 *J. Chem. Phys.* **141** 134314

e-O2

Fragmentation mechanisms for Methane induced by 55 eV, 75 eV and 100 eV electron impact

B. Wei[1]**, X. Wang, Y. Zhang, R. Hutton and Y. Zou**

Applied Ion Beam Physics Laboratory, Fudan University, Key Laboratory of the Ministry of Education
Institute of Modern Physics, Fudan University, Shanghai 200433, China

Synopsis CH_4^{2+} break up into three or more fragments in one or two-step processes were studied, our result shows that there are three reaction channels to form CH_2^+, H^+, and H, one synchronous concerted reaction channel and two two-step reaction channels. For even more complicated fragmentation processes of CH_4^{2+} dications, the fragmentation mechanism can still be identified in the present measurements.

The fragmentation of CH_4^{2+} dications following 55 eV, 75 eV and 100 eV electron impact double ionization of methane was studied using a cold target recoil-ion momentum spectroscopy. From the measured momentum of each recoil ion, the momentum of the neutral particles has been deduced and the kinetic energy release distribution for the different fragmentation channels has been obtained. The doubly charged molecular ions break up into three or more fragments in one or two-step processes, resulting in different signatures in the data. We observed the fragmentation of CH_4^{2+} dications through different mechanisms according to the momentum of the neutral particles. For example, our result shows that there are three reaction channels to form CH_2^+, H^+ and H, one synchronous concerted reaction channel and two two-step reaction channels. For even more complicated fragmentation processes of CH_4^{2+} dications, the fragmentation mechanism can still be identified in the present measurements. The slopes of the peak in the ion-ion coincidence spectra were also estimated here, as they are also related to the fragmentation mechanism.

The neutral fragments produced in the two-step fragmentation shared the momentum with the charged fragments. However, in the synchronous concerted fragmentation, the neutral fragments did not take part in the Coulomb interaction, and thus there is no momentum sharing between the neutral fragments and the charged fragments. To identify the fragmentation mechanism of CH_4^{2+} ions into three bodies, Newton diagram was used in our analysis. Figure 1 show the Newton plot for the CH_4^{2+} diaction dissociating into CH_2^+, H^+ and H induced by 100 eV electrons. The momentum vector of H^+ ion defines the x axis, shown in Figure 1 as an arrow. The relative momentum vectors of CH_2^+ ions and neutral H are mapped in the upper and lower half respectively. The momentum of neutral H ranges from − 20 a.u. to + 20 a.u.. In the fourth quadrant, there are some neutral H atoms with a momentum in the same direction as H^+. While in the third quadrant, there are some neutral H atoms with a momentum in the same direction as CH_2^+. It indicates that neutral H shared momentum with H^+ and CH_2^+, respectively. Same measurement has been performed for CO_2. More detail can be found in Reference [1, 2].

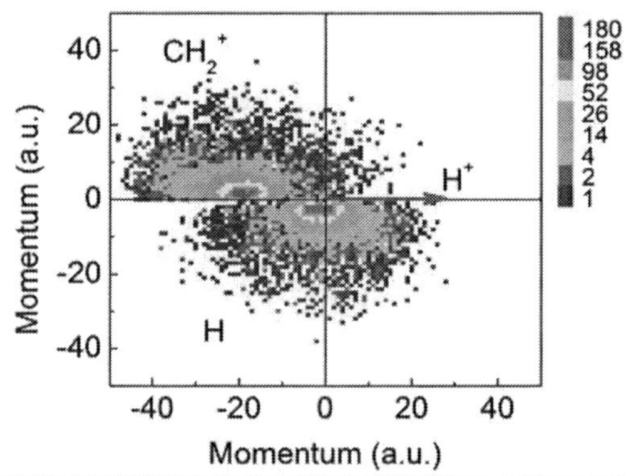

Figure 1. Newton diagram for the Coulomb explosion process of $CH_4^{2+} \rightarrow CH_2^+ + H^+ + H$, which induced by 100 eV electron impact.

References

[1] B. Wei, Y. Zhang, X. Wang, *et al*, *J. Chem. Phys.* **140**, 124303 (2014)
[2] X. Wang, Y. Zhang, D. Lu, *et al*, *Phys. Rev. A*, **90**, 062705 (2014)

[1] E-mail: brwei@fudan.edu.cn

First observation of coherence in a highly charged ion

S. Tashenov*[1], D. Banaś†, H. Beyer‡, C. Brandau♮,♯, S. Fritzsche♭,÷, A. Gumberidze♮,♯, S. Hagmann‡, P.-M. Hillenbrand‡, H. Jörg*, I. Kojouharov‡, Ch. Kozhuharov‡, M. Lestinsky‡, Yu.A. Litvinov*,‡, A.V. Maiorova∅, H. Schaffner‡, V.M. Shabaev∅, U. Spillmann‡, Th. Stöhlker♭,∇, A. Surzhykov♭, S. Trotsenko‡,♭

*Physikalisches Institut der Universität Heidelberg, 69120 Heidelberg, Germany
†Institute of Physics, Jan Kochanowski University, 25-406 Kielce, Poland
‡GSI Helmholtzzentrum für Schwerionenforschung GmbH, 64291 Darmstadt, Germany
♮ExtreMe Matter Institute EMMI and Research Division, GSI Helmholtzzentrum für Schwerionenforschung, 64291 Darmstadt, Germany
♯FIAS Frankfurt Institute for Advanced Studies, 60438 Frankfurt am Main, Germany
♭Helmholtz-Institut Jena, 07743 Jena, Germany
÷Theoretisch-Physikalisches Institut, Universität Jena, 07743 Jena, Germany
∅St. Petersburg State University, 198504 St. Petersburg, Russia
∇Institut für Optik und Quantenelektronik, Universität Jena, 07743 Jena, Germany

Synopsis We report the first observation of correlated x rays emitted by a heavy highly charged ion such as hydrogenlike uranium in the process of radiative recombination. The angular correlations between these x rays for the first time revealed a coherence of the populated state in this ion. The results also indicate a strong contribution of the spin–orbit interaction in radiative recombination.

Radiative recombination (RR) into highly charged ions offers a unique opportunity to study its time-reverse the fundamental process of the photoelectric effect in the otherwise inaccessible regime of hard x rays and strong Coulomb fields. Using the experimental storage ring ESR in GSI Darmstadt we have studied the RR into the excited $2p_{3/2}$ state of hydrogenlike uranium ions. This state decays to the ground state $1s_{1/2}$ by emitting a Lyα_1 x ray. In the experiment the RR and the Lyα_1 x rays were detected in time-coincidences using a setup of large germanium detectors, representing the first observation of the correlated x rays emitted by heavy highly charged ions [1].

We observed that the angular distribution the Lyα_1 x rays is strongly correlated with the emission direction of the RR x ray. To interpret this observation we consider a non statistical population of the magnetic sublevels (alignment) of the $2p_{3/2}$ state. When the RR x ray is not observed, as was the case in all previous alignment experiments with highly charged ions, the populated state is axially symmetric around the collision direction, and, therefore, its magnetic sublevels are populated incoherently. In contrast, in the present experiment, this state is populated *coherently*. Therefore, it loses its symmetry with respect to the collision axis. Figure 1 shows the effect of the coherence on the electric charge density distribution of the $2p_{3/2}$ state. The alignment axis of this distribution attains a finite angle γ with respect to the electron propagation direction. This angle is neither 0° nor 90° which manifests the coherence of the populated state. We extracted the alignment angle of the charge cloud from the measured angular distribution of the Lyα_1 x rays revealing for the first time the coherence in a heavy highly charged ion. Moreover, the experimental results indicate a strong contribution of the spin-orbit interaction to the RR process and, therefore, to the fundamental process of the photoelectric effect.

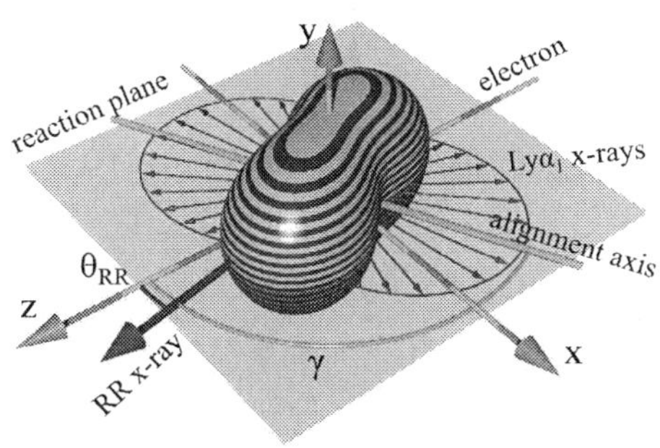

Figure 1. The angular density distribution of the charge cloud of the RR-populated $2p_{3/2}$ state when the RR x ray is observed.

References

[1] S. Tashenov *at al* 2014 *Phys. Rev. Lett.* **113** 113001

[1]E-mail: tashenov@physi.uni-heidelberg.de

e-O4

Iterative calculation of electron energy loss spectra

Peter Koval* [1], **Mathias Per Ljungberg*** [2], **Dietrich Foerster**† [3], **and Daniel Sanchez-Portal*** [4]

* Donostia International Physics Center, San Sebastian, Spain;
* Centro de Fisica de Materiales, San Sebastian, Spain.
† Laboratoire Ondes et Matiére d'Aquitaine, Bordeaux, France.

Synopsis We propose an algorithm of low complexity for the calculation of electron energy loss spectra (EELS) within time-dependent density functional theory (TDDFT). Key ideas of the method include exploiting the locality of the response function, expressing the Kohn-Sham orbitals in terms of atomic orbitals and employing an iterative method to avoid calculating the whole matrix representation of the response functions.

High-energy electrons from a transmission electron microscope can be used to probe the electronic properties of materials by measuring the energy and momentum losses the electrons experience when undergoing inelastic scattering. In the linear response regime the EELS cross section is connected to the dielectric function (DF).

In this contribution we present our method [1] of computing the DF within time-dependent density functional theory and using localized basis functions. The non-interacting response function $\chi^0_{\boldsymbol{G}\boldsymbol{G}'}(\boldsymbol{q},\omega)$ is expressed in terms of Fourier transforms of our localized basis functions $F^{\mu}_{\boldsymbol{G}}(\boldsymbol{q})$

$$\chi^0_{\boldsymbol{G}\boldsymbol{G}'}(\boldsymbol{q},\omega) = F^{\mu}_{\boldsymbol{G}}(\boldsymbol{q})\chi^0_{\mu\nu}(\boldsymbol{q},\omega)\overline{F}^{\nu}_{\boldsymbol{G}'}(\boldsymbol{q}), \qquad (1)$$

where the matrix $\chi^0_{\mu\nu}(\boldsymbol{q},\omega)$ is explicitly known. The interacting response function, expressed analogously in terms of localized functions, obeys the matrix equation $\chi = (1-\chi^0 K)^{-1}\chi_0$ and the observable macroscopic DF is connected to the matrix of the interacting response function $\chi_{\mu\nu}(\boldsymbol{q},\omega)$ by

$$\epsilon^{-1}(\boldsymbol{q},\omega) = 1 + \frac{4\pi}{q^2}F^{\mu}_{\boldsymbol{0}}(\boldsymbol{q})\chi_{\mu\nu}(\boldsymbol{q},\omega)\overline{F}^{\nu}_{\boldsymbol{0}}(\boldsymbol{q}). \qquad (2)$$

The expression of the response functions in the basis of localized functions $\chi^0_{\mu\nu}(\boldsymbol{q},\omega)$ bears an advantage over a more common representation in the basis of plane waves $\chi^0_{\boldsymbol{G}\boldsymbol{G}'}(\boldsymbol{q},\omega)$. Namely, the application of the non-interacting response $\chi^0_{\mu\nu}(\boldsymbol{q},\omega)$ onto a vector can be formulated in terms of sparse matrix-vector products with a complexity no worse than $O(N^3)$ where N is number of atoms in the unit cell. Moreover, an iterative method for computation of macroscopic DF typically requires only few (10–20) applications of non-interacting response and is trivially parallel in frequency ω. The method is suited for the large unit cells characteristic for molecular solids. In the figure, the EELS of fullerite is shown in comparison to experimental spectra.

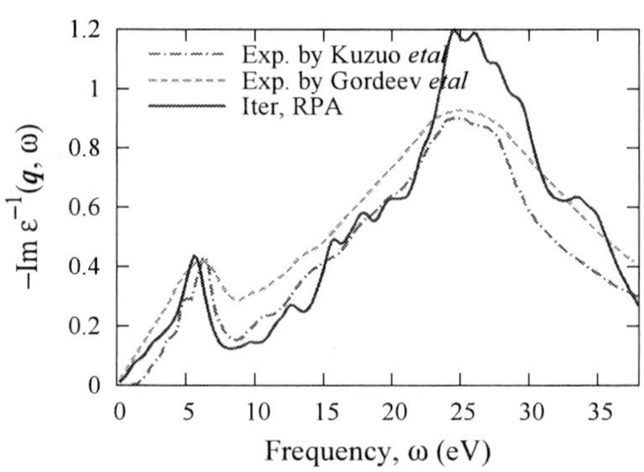

Figure 1. EELS of solid C60 computed with our iterative method is compared to measured spectra. Experimental data is taken from [2] and [3].

References

[1] P. Koval *etal* 2014 *Nucl. Instr. Meth. B* in press `http://dx.doi.org/10.1016/j.nimb.2014.11.080`

[2] R. Kuzuo, *etal* 1991 *Jpn. J. Appl. Phys.* **30** L1817.

[3] Y. Gordeev, *etal* 2000 *Phys. Solid State* **42** 381.

[1]E-mail: koval.peter@gmail.com
[2]E-mail: mathias.ljungberg@gmail.com
[3]E-mail: dietrich.foerster@u-bordeaux.fr
[4]E-mail: sqbsapod@ehu.eus

The ground state nuclear dynamics of dimethyl ether in momentum space

F. Morini* [1]**, N. Watanabe**†**, M. Kojima**†**, M. S. Deleuze* and M. Takahashi**†

*Center of Molecular and Materials Modelling, Hasselt University, Agoralaan Gebouw D, B-3590 Diepenbeek, Belgium

† Institute of Multidisciplinary Research for Advanced Materials, Tohoku University, Sendai 980- 8577, Japan

Synopsis The role of molecular vibrations has been analyzed for the outer valence (e, 2e) momentum profiles of dimethyl ether using two different approaches. The first is cast over the harmonic oscillator model, while the second relies upon the principles of the Born-Oppenheimer Molecular Dynamics (BOMD).

In this study an investigation of the thermally induced nuclear dynamics of the electronic ground state of dimethyl ether is performed on its outer valence orbitals (e, 2e) momentum profiles. To this aim, two different methodologies have been used. The first one, known as Harmonic Oscillator Quantum Mechanical (HAQM) approach [1], allows to quantitatively evaluate the role of each normal mode of vibration on a given momentum profiles. The second one uses the principles of Born-Oppenheimer Molecular Dynamics (BOMD) simulations [2] and allows, by invoking the ergodic principle, to average the effects of the temperature over a large number of structures and to take into account non-harmonic effects.

The obtained theoretical results have been compared with both newly acquired experimental (e, 2e) momentum profiles and previous measurements using a different instrumental set-up [3].

The final outcome shows that, in contrast with what has been stated so far about the role of vibrations for this molecular target [3], the inclusion of vibrational effects in the calculations significantly affects the shape and intensity of the momentum profiles, and thus explains the discrepancies observed between theory and experiment regardless the instrumental parameters (see e. g. figure 1). In line with what has been found by similar studies [4], specific high frequency modes of vibration, which mostly involve the C-H bonds, are the main cause for such variations. The results provide a thorough overview of the interplay existing in this molecule between the electronic and vibrational levels that can be fully accounted for according to the Herzberg-Teller principle, and confirm the consistency of both HAQM and BOMD approach.

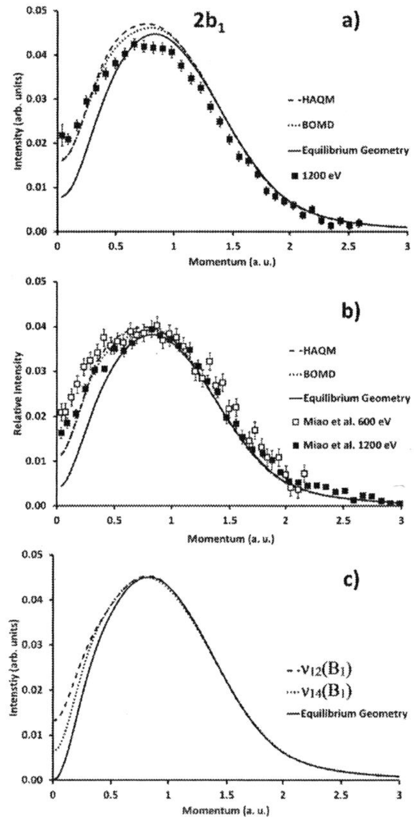

Figure 1. Vibrational effects upon the momentum profiles of the $2b_1$ orbital: a) comparison with the new measurements, b) comparison with ref. 3, and c) contribution of the relevant modes of vibration.

References

[1] N .Watanabe *et al* 2012 *J. Chem. Phys.* **137** 114301

[2] See e. g. S. H. R. Shojaei *et al* 2013 *J. Phys. Chem. A* **117** 1918 and references therein

[3] Miao *et al* 2014 *J. Electron Spectrosc. Relat. Phenom.* **193** 1

[4] See e. g. F. Morini *et al* 2015 *J. Chem. Phys.* **142** 094308 and references therein

[1] E-mail: filippo.morini@uhasselt.be

e-O6

Inelastic e-Cd and e-Zn collisions

M. Piwiński [1], Ł. Kłosowski, D. Dziczek, S. Chwirot

Institute of Physics, Faculty of Physics, Astronomy and Informatics, Nicolaus Copernicus University, Grudziądzka 5, 87-100 Toruń, Poland

Synopsis Coherence analysis version of the electron-photon coincidence technique was applied to study the electron impact excitation of singlet-P state of cadmium and zinc atoms. The experimental results of the Stokes and Electron Impact Coherence Parameters (EICPs) for various collision energies are presented and discussed.

Electron-atom collision is one of the fundamental processes in atomic physics which has been extensively studied since Franck and Hertz published their pioneering work [1]. Despite this fact there are still difficulties in proper description of these phenomena especially for more complex (heavier) atomic targets.

There are various experimental techniques which can deliver information about scattering amplitudes characterizing collision process. However, only few of them are able to provide information not only about moduli but also about relative phases of relevant scattering amplitudes.

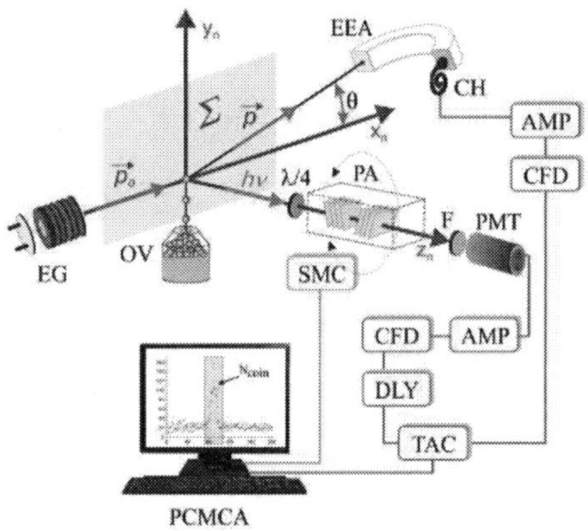

Figure1. Schematic of the apparatus and geometry of the electron-photon coincidence experiment in the coherence analysis version. EG – electron gun, EEA – electron energy analyser with channeltron (CH), OV – the source of the atomic beam, λ/4 – zero-order quartz retardation plate, PA – "pile-of-plates" polarisation analyser, F – optical filter, PMT – photomultiplier tube, CFD – constant-fraction discriminator, SMC – stepper motor controller, TAC – time-to-amplitude converter, AMP – preamplifier, DLY – delay line, PCMCA – personal computer with multichannel analyser.

Such experimental research can be conducted using electron-photon coincidence technique in either coherence analysis [2] or angular correlation version [3]. The Electron Impact Coherence Parameters (EICP) values provide the most complete characterization of the inelastic collision and allow for a verification of proposed theoretical models [4].

The electron-photon coincidence methods deliver such detailed information at a cost of time-consuming measurements lasting from several days to several weeks for a single data point. Therefore experiments of this kind have resulted in rather limited amount of data.

Present work is a continuation of our systematic studies on electron impact excitation of atoms with two valence electrons, from the simplest case He [3,5] to more complicated, heavier elements with closed inner shells: Ca [6,7], Zn [2,8,9] and Cd [10,11,12].

We present the results of Stokes and EICP parameters characterising electronic excitation of the lowest singlet P-state state of cadmium and zinc atoms for various collision energies. The experiment was performed using electron-photon coincidence technique in the coherence analysis version. The obtained data are presented and compared with existing theoretical predictions.

References

[1] J. Franck and G. Hertz 1914 *Verh. Dtsch. Phys. Ges.* **16**, 457

[2] M. Piwiński *et al* 2012 *Phys. Rev. A* **86**, 052706

[3] Ł. Kłosowski *et al* 2007 *Meas. Sci. Technol.* **18**, 3801–3810

[4] N. Andersen *et al* 1988, *Phys. Rep.* **165**, 1–188

[5] Ł. Kłosowski *et al* 2009 *Phys. Rev. A* **80**, 062709

[6] D. Dyl *et al* 1999 *J. Phys. B: At. Mol. Opt. Phys.* **32**, 837–844

[7] D. Dziczek *et al* 1998 *Acta Phys. Polon. A* **93**, 717–721

[8] M. Piwiński *et al* 2013 *Eur. Phys. J.-Spec. Top.* **222**, 2273–2277

[9] M. Piwiński *et al* 2015 in preparation

[10] M. Piwiński *et al* 2002 *J. Phys. B: At. Mol. Opt. Phys.* **35**, 3821–3827

[11] D. Dziczek *et al* 2003 *Acta Phys. Polon. A* **103**, 3–11

[12] M. Piwiński *et al* 2006 *J. Phys. B: At. Mol. Opt. Phys.* **39**, 1945–1953

[1]E-mail: Mariusz.Piwinski@fizyka.umk.pl

Influence of the higher order effects on the polarization and angular distribution of the radiation following electron-impact excitation process

C. Z. Dong, J. Jiang, L.Y. Xie, Z. B. Chen and Z. W. Wu

* Key Laboratory of Atomic and Molecular Physics Functional Material of Gansu Province, College of Physics and Electronic Engineering, Northwest Normal University, Lanzhou 730070, China

Synopsis: The influence of the higher order effects on the polarization and angular distribution of the radiation following electron impact excitation process have been presented on the basis of the calculations using a new developed fully relativistic distorted wave program.

Electron-impact excitation (EIE) is one of the most important atomic processes occured frequently in various laboratory and astrophysical plasmas. From the polarization and angular distribution of decay products following the collision process, valuable information can be obtained for both the collision dynamics and the magnetic sublevel population of the impact excited states. These polarization data become indispensable for the detailed diagnosis of plasma state and the analysis of complex spectra formation mechanism. To satisfy application requirements, various activities have been attempted to compile data and develop databases. However, the accuracy of these available polarization data is often suspects, different calculations using similar methods/codes sometimes differ so much due to the presence of the higher order effects.

On the basis of the well-known GRASP92/2K and RATIP packages [1], a new fully relativistic distorted wave program, named REIE06 has been developed by our group in recent. In this presentation, some selected applications of the program are shown. A special attention has been paid on influences of the higher order effects on the polarization and angular distribution properties of the radiation following EIE process, such as, the effects of intermediate resonance states in an isolated resonance approximation [2], the effects of Breit interactions [3], the effects of *E*1-*M*2 interference [4], the effects of plasma screening [5] and the effects of radiative cascade [6]. We hope these efforts will be helpful for further understanding the contributions from the higher order effects and getting more accurate polarization data.

This work was supported by the National Natural Science Foundation of China (Grant nos. 11274254 and U1332206).

References

[1] P Jonsson, X. He, C.F. Fischer and I. P. Grant, CPC 177, 597 (2007); S. Fritzsche, H. Aksela, and C. Z. Dong et al. Nucl. Instrum. Methods Phys. Res. B 205, 93 (2003)

[2] J. Jiang, C. Z. Dong, L. Y. Xie, and J. G. Wang, Phys. Rev. A 78, 022709 (2008)

[3] Z. W. Wu, J. Jiang, and C. Z. Dong, Phys. Rev. A 84, 032713 (2011)

[4] Z. B. Chen, C. Z. Dong, L. Y. Xie, and J. Jiang, Phys. Rev. A **89**, 022513 (2014)

[5] B. W. Li, C. Z. Dong, J. Jiang, and J. G. Wang, Eur. Phys. J. D 59 201(2010)

[6] Z. B. Chen, C. Z. Dong and J. Jiang, Phys. Script (2015) in press

E-mail: dongcz@nwnu.edu.cn

e-O8

Projectile coherence – the transition between the semiclassical and the quantum method

L. Nagy[*1], **F. Járai-Szabó**[*]

[*] Faculty of Physics, Babeş-Bolyai University, Kogălniceanu Street No. 1, 400084 Cluj, Romania

Synopsis In the present work fully differential cross sections for the ionization of helium by fast charged projectiles are calculated by different methods. Beside a semiclassical model, assuming a straight-line trajectory for the projectile, and the quantum model, considering it a plane wave, we present a more realistic method, describing the projectile as a wave packet with a given coherence width. By these calculations we study the effect of projectile coherence on the cross sections.

When precise and complete measurement of the fully differential cross section (FDCS) for the ionization of the helium has been performed [1], it has been observed significant difference between the experimental data and theoretical continuum distorted wave (CDW) results. These, and other similar theoretical calculations were not able to reproduce the experimentally observed structures in the perpendicular plane. However, our semiclassical calculations [2] and the convolution of the first Born approximation with elastic scattering [3] described fairly well these structures. The discrepancy between these calculations and there relationship with the experimental data was solved, when it has been shown the importance of the projectile coherence in the details of the FDCs [4]. There have been obtained significant differences in the FDCSs for the ionization of helium, using projectiles causing the same perturbation, but different transversal coherence lengths. If the coherence length was larger than the atomic dimensions, the results were closer to the CDW calculations, which assumes a plane wave for the projectile, while for smaller coherence lengths the results were closer to the convoluted [3] and our semiclassical results. In the experiments published in 2003 [1] the projectiles were incoherent (with small coherence length), and this was the reason for the discrepancy with the CDW calculations, describing the projectile as a plane wave.

In a recent paper we have shown the dependence of the FDCS on the projectile coherence [5]. Here, starting from the same transition amplitude, we have calculated the FDCS in the two extreme cases: semiclassically, assuming a straight-line trajectory for the projectile, and using the full quantum method, considering the projectile a plane wave.

Now we complete our previous work with a more realistic description of the projectile: a wave packet with a given, finite coherence width. In the experimental studies the transversal coherence lengths vary from 10^{-3} a.u. to 4 a.u. [4]. In the present work we investigate, how the transition is between the totally incoherent and the totally coherent projectiles, and identify the main effects of the projectile coherence on the TDCSs.

This work was supported by a grant of the Romanian National Authority for Scientific Research (CNCS UEFIS-CDI) project number PN-II-ID-PCE-2011-3-0192.

References

[1] Schultz M *et* al. 2003 *N*ature **422** 48.

[2] F. Járai-Szabó, L.Nagy 2009 *N*ucl. Instrum. Methods B **267** 292.

[3] M. Schulz, M. Dürr, B. Najjari, R. Moshammer and J. Ullrich 2007 *P*hys. Rev. A **76** 032712.

[4] Wang X *et* al. 2012 *J.* Phys. B: At. Mol. Opt. Phys. **45** 211001.

[5] F. Járai-Szabó and L. Nagy 2015 *E*ur. Phys. J. D **69** 4.

[1]E-mail: lnagy@phys.ubbcluj.ro

e-O9

Electron-electron orbital entanglement in two-electron ions around the critical charge region

Yew Kam Ho* [1], **Chien-Hao Lin*** [2]

* Institute of Atomic and Molecular Sciences, Academia Sinica, Taipei, Taiwan

Synopsis Quantum entanglement for two-electron ions is investigated in the region around the critical charge when the ion starts to become unbound and turns into a shape resonance. The electron-electron orbital entanglement entropies (linear and von Neumann) for the shape resonances are quantified by adopting the Schmidt decomposition method with resonance wave functions obtained with the help of complex scaling method.

Recently, we have studied the linear and von Neumann entropies of the ground and singly excited states in helium and the ground states of hydrogen negative ion and positronium ion [1]. Here, we extend our investigation to the two-electron ions near the region of critical charge, Z_{cr}, below which the three-particle system becomes unbound [2], and turns into a shape resonance lying above the ground state of the one-electron atom with charge Z. For S-states we use Hylleraas-type wave functions to take into account of correlation effects, with

$$\Psi_{kmn}(\mathbf{r}_1,\mathbf{r}_2)=\sum_{kmn} C_{kmn}\left\{\exp\left[-\alpha\left(r_1+r_2\right)\right]r_1^k r_2^m r_{12}^n+\left(1\leftrightarrow 2\right)\right\},$$

where 1 and 2 denotes the electron 1 and electron 2, respectively, and r_{12} is the relative distance between the electron 1 and electron 2, and k, m, n are positive integers or zero. Up to N=444 terms are used. When $Z > Z_{cr}$, the state remains bound and their entropies are obtained by using the energy optimized (minimized) wave functions. For the region when $Z < Z_{cr}$, we employ the complex scaling method [3] to determine the resonance energy and width. We next adjust the non-linear parameters in the wave function to have the energy match the real part of the complex pole. Once the wave function for a given Z is obtained, it is used to quantify linear entropy and von Neumann entropy by using the Schmidt decomposition method [1]. Figures 1 and 2, respectively, show the shape resonance energy and its half-width *vs* $1/Z$. Figure 3 shows the von Neumann and linear entropies *vs* Z in the region of critical charge.

Acknowledgement

This work is supported by MOST of Taiwan.

References

[1] C.-H. Lin, Y.-C. Lin, Y. K. Ho, Y. K. 2013 *Few-Body Systems* **54** 2147; C.-H. Lin and Y. K. Ho 2014 *Few-Body Syst.* **55** 1141; C.-H. Lin and Y. K. Ho 2014 *Phys. Lett. A* **378** 2861

[2] C.S. Estienne *et al* 2014 *PRL* **112** 173001.

[3] Y. K. Ho, 1983 *Phys. Reports* **99** 1

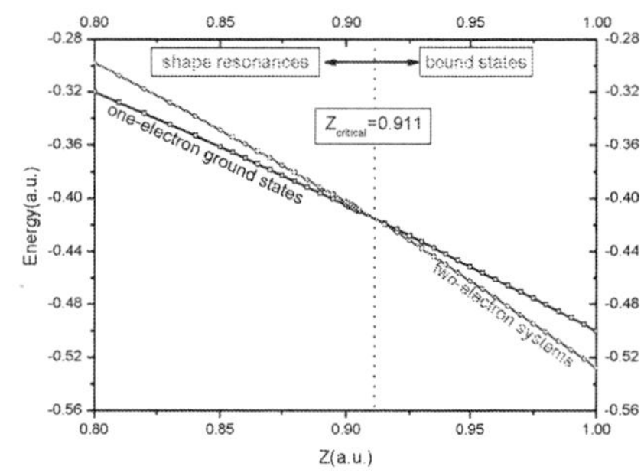

Figure 1. Shape resonance energy *vs* Z, as related to the one-electron ground state energy.

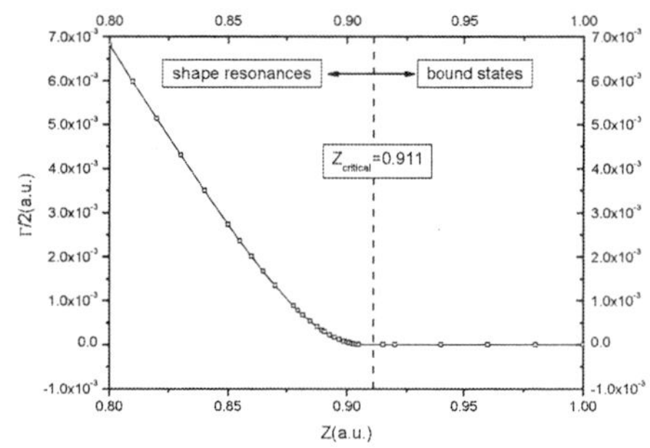

Figure 2. Half-width of shape resonance *vs* Z.

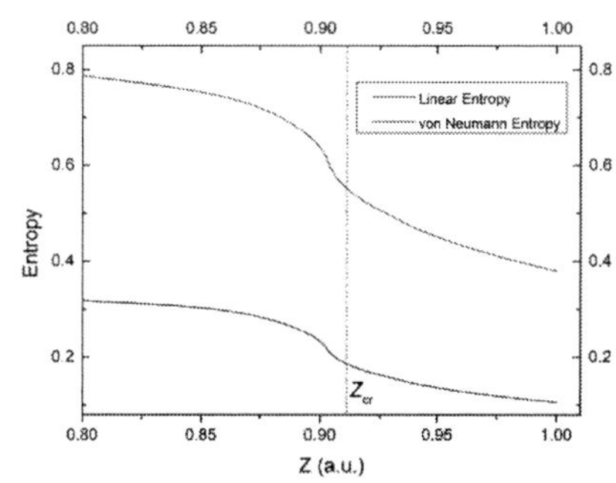

Figure 3. Linear and von Neumann entropies for the two-electron ions.

[1] E-mail: ykho@pub.iams.sinica.edu.tw

e-O10

Effects of the sudden electron-hole pair creation and of the life time of the residual hole on plasmon excitations in surfaces

Juana L. Gervasoni* [1] and **L. Kövér** [† 2]

*Centro Atomico Bariloche and Instituto Balseiro, Comision Nacional de Energia Atomica, S. C. de Bariloche, Rio Negro, Argentina. Member of the Consejo Nacional de Investigaciones Cientificas y Tecnicas (CONICET), Argentina.
† Institute for Nuclear Research of the HAS (MTA ATOMKI), 18/c Bem ter, H-4026 Debrecen, Hungary.

Synopsis We study the process of excitation of bulk and surface plasmons during the emission of electrons in the proximity of a solid surface. We take into account the effects due to the sudden creation of an electron and the residual holes, one in the case of X-ray photoemission spectroscopy (XPS) and two in the case of Auger electron spectroscopy (AES).

The use of systems of spectroscopy and analysis of surfaces have had an important impact in the development of this area [1]. The various types of electron spectroscopies can be divided in two groups. In the first, the surface is bombarded with an electron beam and the scattered (or emitted) electrons are analyzed in angle or energy (secondary electrons). A very popular technique included inside this subgroup is the Auger Electron Spectroscopy (AES). In the second group of electron spectroscopy, they are produced by means of the irradiation of the material with photons generated by an external source. This technique has attracted great interest, stimulated by the possibility of having continuous sources of synchrotron radiation. Nevertheless, most experiments are still made with conventional non-continuous sources. This is the reason why we can distinguish X-ray photoemission spectroscopy (XPS) for fixed energies in the range of KeV [1].

Although these techniques of electron spectroscopy are used with success in surface physics, we must take into account that the use of electrons as carriers of information about surfaces has some inherent limitations. One of these comes from the definition of the surface that we want to study. In this work we study the effect of the sudden electron-hole pair creation and of the life time of the residual hole on plasmon excitations in surfaces, for both processes: XPS and Auger, and its corresponding plasmon excitations.

[1] E-mail: gervason@cab.cnea.gov.ar
[2] E-mail: lkover@esca.atomki.hu

We treat the electron emission in a very simplified manner in order to be able to develop analytical expressions accounting for surface and bulk plasmon excitations. A detailed description of the photoelectron process was done in ref. [2, 3]. Regarding the electron emission, it will be schematized as a sudden creation of a static positive charge a= +e (XPS); +2e (AES) and a moving negative charge $q_- = -e$, which is located at the same site as the positive charge at the moment of creation (See Figure 1). But now the hole has a given life time, given place to some novel behavior that we study in detail.

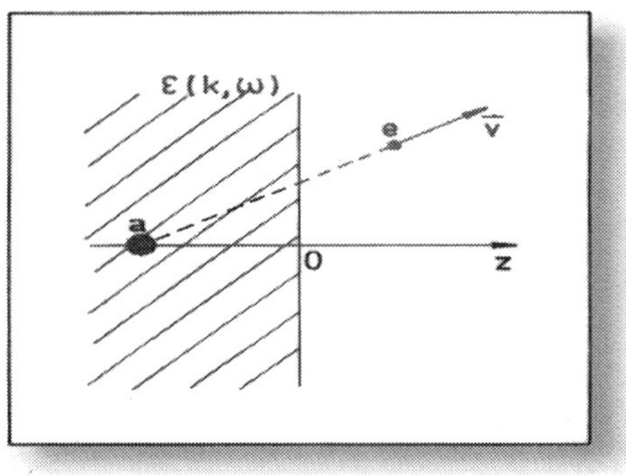

Figure 1. Squeme of the system. We consider the emission of the electron **e** with velocity **v**, from a depth $\mathbf{z_0}$ inside the surface. The hole **a** is created in $\mathbf{z_0}$.

References

[1] H. Ibach, Electron Spectroscopy for Surface Analysis, Springer-Verlag, Berlin, (1977).
[2] J. L. Gervasoni and N. R. Arista, *Surf. Sci.* 260, 329 (1992).
[3] J. L. Gervasoni and L. Kovér, Vacuum, Vol. 83, pp. 1049-1053 (2009).

Optical Constants of Iron Derived from Reflection Electron Energy-Loss Spectra

H. Xu*, B. Da*, S.F. Mao‡, J. Tóth†, K. Tőkési†[1], and Z.J. Ding*[2]

*Hefei National Laboratory for Physical Sciences at Microscale and Department of Physics, University of Science and Technology of China, Hefei, Anhui 230026, P.R. China
‡ School of Nuclear Science and Technology, University of Science and Technology of China, Hefei, Anhui 230026, P.R. China
† Institute for Nuclear Research, Hungarian Academy of Sciences (ATOMKI), Debrecen, Hungary, EU

Synopsis Reflection electron energy loss spectra of iron were measured from 0.5 keV up to 5 keV primary energies and in a wide energy-loss range. An improved reverse Monte Carlo simulation for determination of optical constants via accurate description of electron inelastic transport process of these spectra was performed. The energy loss function of iron extracted from experimental spectra was obtained. The accuracy of the obtained optical data is confirmed by f-sum and ps-sum rules.

Precise value of electron inelastic mean free path (IMFP) is necessary for performing a quantitative analysis by modern surface electron spectroscopy techniques. On the other hand theoretical determination of IMFP relies on optical constants of solid. However, accurate measurement of optical data by optical methods in a photon energy range up to 10^2 eV is still insufficient. Fortunately, such information of optical constants is essentially contained in and, therefore, may be extracted from a reflection electron energy loss spectroscopy (REELS) spectrum due to the shorter information depth of signal electrons compared with that of photons.

In this work we present an improved reverse Monte Carlo (RMC) method for extracting optical constants [1]. The principle of this method is to utilize an iterative process of global searching oscillator parameters modeling optical data by comparing Monte Carlo simulated REELS spectra with experimental measurements. In each iteration step, the dielectric function $\varepsilon(\omega,q)$ or energy loss function (ELF) $\mathrm{Im}\{-1/\varepsilon(\omega,q)\}$ is used as an input for modeling electron inelastic scattering via both bulk excitation and surface excitation. In an iteration step the oscillator parameters are modified taking into account of the difference between the simulated and measured REELS spectra, according to a Markov chain Monte Carlo (MCMC) sampling procedure.

Here we employ the Metropolis-Hastings algorithm in a simulated annealing method for global searching of optimal ELF in hyper parameter space. We improved the modeling of electron inelastic scattering by adopting a depth-dependent semi-classical electron inelastic scattering differential cross section in the surface region. Contributions from surface and bulk electronic excitations and multiple electron scattering effects are also included in a Monte Carlo simulation of REELS spectrum.

Measurements of REELS spectra of iron have been carried out for electron primary energies of 0.5, 1, 2, 3, 4, and 5 keV in a wide energy loss range of 0-200 eV. Figure 1 displays the ELFs extracted by the present method (red curve) with error range indicated by a gray area. Other data from atomic photo-absorption experiments (black dots), DFT calculation (yellow squares) and a deconvolution method (blue squares) are also displayed for a comparison.

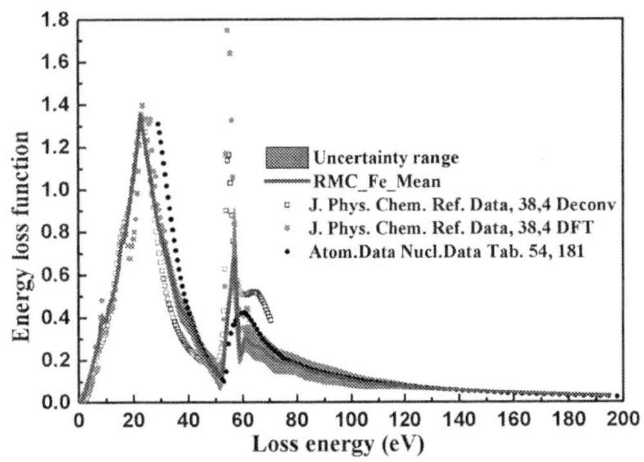

Figure 1. The energy loss functions of Fe.

This work was supported by National Natural Science Foundation of China (Nos. 11274288 and 11204289), National Basic Research Program of China (Nos. 2011CB932801 and 2012CB933702), Ministry of Education of China (No. 20123402110034), the Hungarian Scientific Research Fund (OTKA Nos. NN 103279 and K103917) and COST Action CM1204 (XLIC). We thank the Supercomputing Center of USTC for support in performing parallel computations.

References

[1] B. Da *et al* 2013 *J. Appl. Phys.* **113** 214303.

[1] e-mail: tokesi@atomki.mta.hu
[2] e-mail: zjding@ustc.edu.cn

ph-O1

One photon-two electron ionization in atoms and endohedrals – simplicity and complexity

M. Ya. Amusia* [1, **], **E. G. Drukarev**[† 2]

[*]Racah Institute of physics, The Hebrew University, Jerusalem, Israel
[**]Ioffe Physical-Technical Institute, St. Petersburg, Russia
[†]Konstantinov Petersburg Nuclear Physics Institute, Gatchina, Russia

Synopsis We analyze the mechanisms of one photon-two electron ionization in atoms and multi-atomic formation. The role of target's multiple ionization proved to be decisively important.

One can distinguish three mechanisms of one photon-two electron interactions in a bound system if the photon energies are large enough. In the *shake-off* (SO) the photon knocks out a bound electron, while the other can move to an excited state due to alteration of the charge of the system.

Final-state interaction of the ejected electron with the bound system (FSI) is an alternative mechanism. In both cases single photoionization is the first step, and the mechanisms require large recoil momentum. The photon can interact also with a correlated pair of the bound electrons with small recoil momentum, providing the *quasi-free mechanism* (QFM). In atoms all three mechanisms are important at different photon energies and kinematics.

We focus on multiple photoionization of the endohedral atoms, in which the photon interacts directly with the caged atom, while the other electrons come from the fullerene shell (FS). Since the FS radius is big $R >> 1$, and thus is much larger than the size of the caged atom, the SO amplitude is suppressed by a factor $1/R$. The QFM proceeds at small distances between the bound electrons, and is not important in this case. The FSI interaction with *each* electron of the FS is suppressed by a factor of $1/E$, with E being the photoelectron energy. However, there are $N >> 1$ electrons in the FS, and the FSI contribution requires additional analysis. Due to FS large size it became possible to sum the perturbative series for the FSI including any final state of the FS.

We found the total probability P of inelastic processes in the FS during the photoionization of the caged atom employing the closure of the FS functions. For C_{60} our analysis is true for $E >> 30eV$, when most important excited states can be included. Due to the small thickness $\Delta \sim 1 << R$ of the FS, we obtain the analytical expression $P = 1 - \exp(-N/E)$. Thus, for example, for the fullerene C_{60} we find $P \approx 1$ if $E < 1keV$. For the energies of several keV the FSI contribution drops as $1/E$, and the inelastic processes are the result of interplay between the FSI and the SO.

Our results are confirmed by recent experiments on photoionization of Ce and Xe encapsulated in C_{82} and C_{60}. The probability of photoionization of the caged atom with the FS being unchanged appeared to be much smaller than expected [1]. The results are in line with our calculations [2]. An important consequence of this analysis is that one cannot describe interaction of the photoelectron with the FS by a static potential. The approach developed in our calculations can be employed for investigation of photoionization in other systems, which have two scales of distances. These can be heavy atoms or complex molecules.

In total, the double photoionization can proceed as the single one, followed by the interaction of the outgoing electron with the electronic shell-final state interaction (FSI). In double photoionization of helium it provides noticeable contribution to cross section at the energies of several hundred eV, preceding the asymptotic regime. The FSI mechanism manifests itself brightly in photoionization of endohedral atoms. Due to the large size of the fullerene shell its interactions with the electron knocked out from the caged atom can be calculated beyond the perturbative approach. In the broad interval of energies the probability of inelastic processes in the fullerene shell was found to be close to unity.

Referencies

[1] R. Phaneuf *et al* 2013 *Phys. Rev.* A**88** 053402
[2] M. Amusia *et al* 2014 *JETPLett.* **100** 619

[1] E-mail: amusia@012.net.il

[2] E-mail: edrukarev@list.ru

Single- and double emission from multi-electron systems: How to compute IR double emission and solve the CO_2 mystery

A. Zielinski, V. P. Majety, and A. Scrinzi[1]

Department of Physics, Ludwig Maximilians University, Munich, Germany

Synopsis By a new *ab initio* approach we solve two long-standing problems in strong-field multi-electron systems: fully differential double-emission spectra at IR wave length, where we quantitatively reproduce the measured e-2e ratio and provide an experimentally accessible map of correlated vs. uncorrelated emission in the two-electron energy plane, and the notorious discrepancy between experiment and theoretical prediction for the alignment dependence of strong field CO_2 ionization, which is dominated by to dynamical exchange.

Using a combination of newly developed methods [1, 2, 3, 4] we solved several key problems in strong laser-matter interaction such as (1) the computation of precision photo-electron spectra at elliptic polarization (attoclock, [5, 6]), (2) accurate fully differential (angle-resolved) double emission spectra from the He atom and from H_2 with polarization parallel to the molecular axis, (3) single photo-emission from the CO_2 molecule in dependence of molecular orientation. In all cases we can provide dependable accuracy estimates.

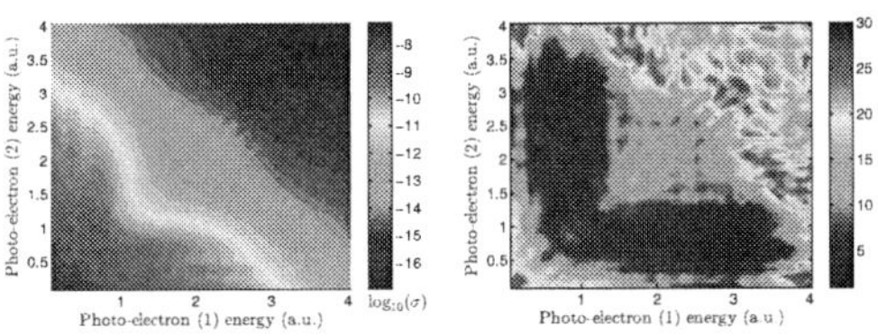

Figure 1. Double emission spectra of He by a 3 cycle 700nm pulse at intensity $3 \times 10^{14} W/cm^2$ (left) and angular correlation (right).

Fig. 1 shows near-IR double emission calculations with accuracies of $\lesssim 5\%$ together with energy-dependent correlation. The SVD-based correlation measure allows a systematic experimental distinction of correlated from uncorrelated emission. Fig. 2 shows the first *ab initio* computational reproduction of the single to double ionization yield ratio at IR wave length.

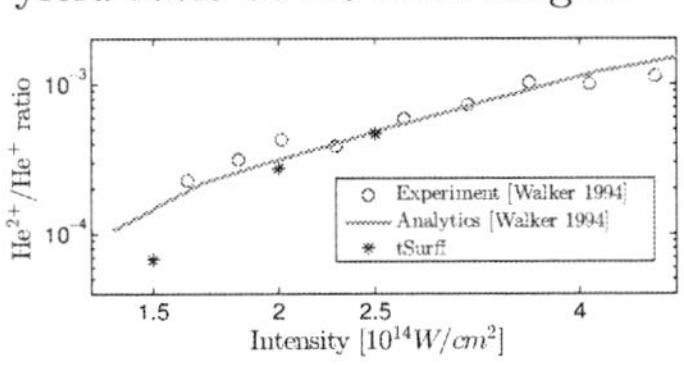

Figure 1. e-2e ratio at 780 nm wave length, experiment (red), fit (green) and *ab initio* (tSurff, blue).

In the haCC approach [4] for computing emission from molecules, strong field dynamics is combined with quantum chemical structure (COLUMBUS, [7]) including full interchannel coupling and exchange. We find that including exchange is crucial: it explains the observed orientation dependence of ionization of CO_2. Multielectron effects can clearly be detected, but have no influence on the peak emission angle at 45°. Fig. 3 shows our results and explains the smaller angles reported in literature.

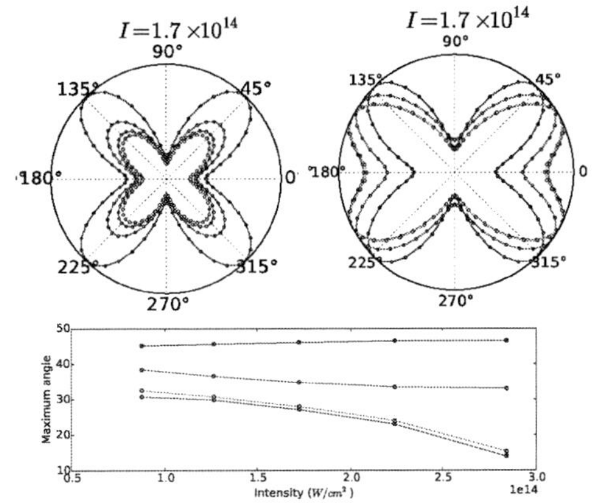

Figure 1. Alignment-dependence of field ionization of CO_2: multi-electron effects reduce the overall yield, but have no influence on angle of peak emission (blue to cyan, left top panel). Lack of anti-symmetrization produces the smaller angles reported in literature (right top panel, red and green). In the full calculation (blue) the emission angle is largely independent of intensity (lower panel).

References

[1] A. Scrinzi. *PRA*, 81:1, 2010.

[2] L. Tao, A. Scrinzi. *New J. Phys.*, 14:013021, 2012.

[3] A. Scrinzi. *New J. Phys.*, 14(8):085008, 2012.

[4] V. P. Majety, A. Zielinski, and A. Scrinzi. *arXiv:1412.3666 [physics.comp-ph]*, 2014.

[5] C. Hofmann et al., *PRA*, 90(4), 2014.

[6] L. Torlina et al. Nature Physics. accepted for pub.

[7] H. Lischka et al. COLUMBUS. *Wiley Interdiscip. Rev.-Comp. Mol. Science*, 1(2):191, 2011.

[1]E-mail: armin.scrinzi@lmu.de

ph-O3

New look at the strong field approximation in laser-matter interactions

A. Galstyan[a] [1], O. Chuluunbaatar[b,c], A. Hamido[a], Yu.V. Popov[d,b], F. Mota-Furtado[e], P. F. O'Mahony,[e] N. Janssens[a] and B. Piraux[a]

[a]Institute of Condensed Matter and Nanosciences, Université catholique de Louvain, B-1348 Louvain-la-Neuve, Belgium

[b]Joint Institute for Nuclear Research, Dubna, Moscow region 141980, Russia

[c] School of Mathematics and Computer Science, National University of Mongolia, UlaanBaatar, Mongolia

[d] Nuclear Physics Institute, Moscow State University, Moscow, 119991, Russia

[e] Department of Maths, Royal Holloway, University of London, Egham, Surrey TW20 0EX, United Kingdom

Synopsis We analyze the validity of the strong field approximation taking into account its dependence on the gauge. We show that the wave packet must be properly normalized to get the results that can be compared to the experimental data. We introduce and discuss a fully gauge invariant Born series in the Coulomb potential.

Strong field approximation (SFA) is used in laser-atom physics for about 50 years. The formulation of the approximation exists for both velocity (V-) and length (L-) gauge. However, despite the fact that the results are different in both gauges [1] both formulations are assumed to be valid. In this contribution we discuss this problem and introduce a correct, fully gauge invariant Born series.

The SFA is based on the assumption that after the atomic electron is ejected, it is essentially driven by the external electric field while the Coulomb potential can be treated as a perturbation. Starting from the time dependent Schrödinger equation (TDSE) in the V-gauge for the atomic hydrogen ($(1/c)A(t) = -b'(t)$),

$$\left[i\frac{\partial}{\partial t} + \frac{1}{2}\triangle_r + \frac{Z}{r} - ib'(t)(\vec{e}\cdot\vec{\nabla}_r)\right]\Phi_V(\vec{r},t) = 0, \tag{1}$$

with $\Phi_V(\vec{r},0) = \varphi_0(r)$, it is possible to derive a fully gauge invariant Born series in the Coulomb potential from the corresponding Lippmann-Schwinger equation

$$\Phi_V(\vec{r},t) = i\int d^3r' G_V(\vec{r},t;\vec{r}\,',0)\varphi_0(\vec{r}\,') - Z\int_0^t dt' \int \frac{d^3r'}{r'} G_V(\vec{r},t;\vec{r}\,',t')\Phi_V(\vec{r}\,',t'). \tag{2}$$

Here, the Green's function writes

$$G_V(\vec{r},t;\vec{r}\,',t') = -i\theta(t-t') \times \int \frac{d^3p}{(2\pi)^3}\chi_V(\vec{r},\vec{p},t)\chi_V^*(\vec{r}\,',\vec{p},t'), \tag{3}$$

where $\chi_V(\vec{r},\vec{p},t) = \exp[i\vec{p}\cdot\vec{r} - i(p^2/2)t + ib(t)(\vec{e}\cdot\vec{p})]$ is a Volkov wave.

If we take the first term of the r.h.s of eq. (2) as the starting (zeroth order) term for the Born series, we obtain a gauge invariant expression for each order. We remind the relation between V and L wave packets

$$\Phi_L(\vec{r},t) = e^{-ib'(t)(\vec{e}\cdot\vec{r}) - i\zeta(t)}\tilde{\Phi}_V(\vec{r},t), \tag{4}$$

and $\zeta(t) = (1/2)\int_0^t d\xi\ [b'(\xi)]^2$. This relation is also valid for corresponding Volkov waves.

Problems appear if we introduce an ansatz

$$\Phi_V(\vec{r},t) = e^{-i\epsilon_0 t}\varphi_0(r) + F_V(\vec{r},t). \tag{5}$$

Inserting (5) into (1), we obtain the non-homogeneous TDSE for F_V. Neglecting the Coulomb potential in its l.h.s. we obtain its traditional SFA solution [1] in the V-gauge.

If we insert the wave packet (4) into (1), using again the ansatz (5) and remembering that the electric field $\vec{\mathcal{E}}(t) = b''(t)\vec{e}$, we obtain the SFA solution in the L-gauge. The famous Keldysh formula follows from it [2]. Nevertheless, both V- and L- SFA expressions are not gauge invariant. This means in particular that the function $\exp(-i\epsilon_0 t)\varphi_0(r)$ cannot be simultaneously the starting point for L- and V-gauge invariant Born series.

In the report we discuss in details, why eq. (2) can not be taken like the basic one for the Born decomposition and how to reach LV-gauge invariance with use of (5).

References

[1] D. Bauer *et al.*, Physical Review A, 72 (2005), 023415.

[2] L. V. Keldysh, Sov. Phys. JETP, 20 (1965), 1307.

[1]E-mail: alexander.galstyan@uclouvain.be

Multi-step ultrafast fragmentation of the third row hydrides following *K*-shell excitation and ionisation

Oksana Travnikova[*◇1], **Tatiana Marchenko**[*◇], **Gildas Goldsztejn**[*◇], **Renaud Guillemin**[*◇], **Loïc Journel**[*◇], **Denis Céolin**[‡], **Ralph Püttner**[†], **Hiroshi Iwayama**[∪], **Eiji Shigemasa**[∪], **Maria Novella Piancastelli**[*◇≀] **and Marc Simon**[*◇‡]

[*] CNRS, UMR 7614, Laboratoire de Chimie Physique-Matière et Rayonnement, F-75005 Paris, France
[◇] Sorbonne Universités, UPMC Univ Paris 06, UMR 7614, LCPMR, F-75005 Paris, France.
[†] Institut für Experimentalphysik, Freie Universität Berlin, D-14195 Berlin, Germany
[‡] Synchrotron SOLEIL, l'Orme des Merisiers, Saint-Aubin, F-91192 Gif-sur-Yvette Cedex, France
[≀] Department of Physics and Astronomy, Uppsala University, SE-75120 Uppsala, Sweden
[∪] Ultraviolet Synchrotron Orbital Radiation Facility, Institute for Molecular Science, Okazaki 444-8585, Japan

Synopsis Relaxation dynamics of HCl and H_2S following photoexcitation and photoionisation in the vicinity of the Cl and S *K*-shell thresholds (~2.8 and 2.5 keV, respectively) were studied by recording the Auger LVV decay channel, which is a 2nd step in the relaxation process of the initially created deep core-hole states. Ultrafast dissociation is observed in three well-distinguishable LVV Auger decay channels for both HCl and H_2S following $1s{\rightarrow}LUMO$ excitation. *K*-shell photoionisation of H_2S, in contrast to HCl, also leads to fragmentation on femtosecond timescale due to the dissociative nature of the $S2p^{-2}$ double core-hole state created by the 1st step KLL Auger decay.

The photoexcited or photoionised molecules with deep core holes are highly unstable and short-lived. Their lifetime typically counts only about 1 femtosecond or few hundreds of attoseconds. These deep core-hole states relax by emitting a photon (radiative decay), an electron (non-radiative or *Auger* decay) or often several *Auger* electrons in a cascade multi-step process, thus producing multiply charged ions, which fragment efficiently and rapidly. Previous mass spectroscopy studies on the photo-fragmentation of the 3rd row hydrides (HCl, H_2S) following Cl and S $1s \rightarrow LUMO$ excitations showed that the molecular HCl^{n+}, H_2S^{n+} ion yields are less than 1% and charge states of Cl^{n+} and S^{n+} fragments up to n=6 and n=5 (respectively) were monitored [1, 2]. Moreover, in the same studies of Hansen *et al* [1, 2] a decay channel leading to neutral H-radical ejection with efficiency of 40 and 30% was revealed for the photon excitation energies, matching the top of the broad Cl and S $1s \rightarrow LUMO$ resonance, respectively. Ultrafast dissociation (UFD) leading to the ejection of neutral radicals was previously observed and studied using soft X-ray electron and coincidence spectroscopies in a number of molecular systems for the shallower core-hole states (e.g. Cl2*p*, Br3*d*, S2*p*), whose lifetime is in the range of 3-8 fs [3, 4], but was not seen within the lifetimes of the Cl1*s* or S1*s* core holes (~1 fs). Nuclear dynamics following Cl1*s* $\rightarrow$ *LUMO* excitation was later detected by resonant x-ray emission measurements of the free HCl molecule [5]. However, such a short lifetime of the Cl1*s* core hole (~1 fs) was found to be too short to lead to the complete H-Cl bond breakage and UFD.

In our work we show that UFD occurs following Cl and S *K*-shell excitations not in the 1st but in the 2nd step (i.e. *LVV-Auger*) of the relaxation decay process ~3, 8 and 10 fs after creation of the 1*s* core holes. This is due to the intermediate states with one and two holes in 2*p* core orbitals created after Kα and *KLL-Auger* decays, respectively. UFD is observed for all three possible *LVV-Auger* decays in both HCl and H_2S molecules. Remarkably, UFD is also observed for one *LVV-Auger* decay channel for the excitation energies far above S1*s* ionisation threshold. The dissociative nature of the intermediate $S2p^{-2}$ double core-hole state of H_2S^{2+} is confirmed by theoretical calculations.

Ultrafast nuclear dynamics induced by hard X-rays is predicted to be a rather general phenomenon for larger S-containing polyatomic molecules and it might be of importance in the radiation damage of biomolecules.

[1]E-mail: oksana.travnikova@upmc.fr

References

[1] D. L. Hansen *at al* 1998 *Phys. Rev. A* **57** 2608

[2] D. L. Hansen *at al* 1998 *Phys. Rev. A* **58** 3757

[3] O. Björneholm *et al* 1997 *Phys. Rev. Lett.* **79** 3150

[4] O. Travnikova *et al* 2013 *J. Phys. Chem. Lett.* **4** 2361

[5] M. Simon *et al* 2006 *Phys. Rev. A* **73** 020706(R)

ph-O5

Ultrafast electron dynamics in phenylalanine initiated by attosecond pulses

F. Calegari*, D. Ayuso[†], A. Trabattoni[‡], L. Belshaw[¶], S. De Camillis[¶], S. Anumula[‡], F. Frassetto[§], L. Poletto[§], A. Palacios[†], P. Decleva[⊥], J. Greenwood[¶], F. Martín[†∇] and M. Nisoli[*‡]

[*] Institute of Photonics and Nanotechnologies, IFN-CNR, Piazza Leonardo da Vinci 32, 20133 Milano, Italy
[†] Departamento de Química, Módulo 13, Universidad Autónoma de Madrid, Cantoblanco 28049 Madrid, Spain
[‡] Department of Physics, Politecnico di Milano, Piazza Leonardo da Vinci 32, 20133 Milano, Italy
[¶] Centre for Plasma Physics, School of Maths and Physics, Queen's University Belfast, BT7 1NN, UK
[§] Institute of Photonics and Nanotechnologies, IFN-CNR, Via Trasea 7, 35131 Padova, Italy
[⊥] Dipartimento di Scienze Chimiche e Farmaceutiche, Università di Trieste, 34127, and CNR-IOM, Trieste, Italy
[∇] Instituto Madrileño de Estudios Avanzados en Nanociencia, Cantoblanco, 28049 Madrid, Spain

Synopsis We report the application of isolated attosecond pulses to prompt ionization of the amino acid phenyl-alanine and the subsequent detection of ultrafast dynamics on a sub−4.5-femtosecond temporal scale, which is shorter than the vibrational response of the molecule and can only be associated to purely electronic processes. Our interpretation is confirmed by the good agreement with numerical simulations in which the electronic wave packet generated by the attosecond pulse is quantitatively determined [1].

The process of electron transfer in complex molecules is of crucial importance since it triggers the first steps in a number of biological processes such as photosynthesis, cellular respiration or DNA damage [2]. Theoretical studies have shown that very efficient charge dynamics can be driven by purely electronic effects, which precede any rearrangement of the nuclear skeleton and can evolve on a temporal scale ranging from few femtoseconds down to tens of attoseconds [3,4]. This phenomenon has been referred to as charge migration [3].

Here we present an experimental demonstration of charge migration in a biologically relevant molecule, the amino acid phenylalanine [1]. Charge dynamics was initiated upon ionization by isolated sub-300-as pulses, with photon energy in the spectral range between 15 eV and 35 eV and probed by 4-fs, waveform-controlled near infrared (NIR) pulses, with central wavelength of 720 nm. The parent and fragment ions produced by the interaction with the pulses were collected by a linear time-of-flight device for mass analysis. The fragmentation yield of doubly charged immonium ion as a function of the time delay between the two pulses shows a clear oscillatory pattern which has been fitted to a sinusoidal function of period 4.3 fs. This ultrafast dynamics can only be associated with purely electronic effects, thus constituting the first experimental measurement of charge migration in a biomolecule.

We have performed theoretical calculations to describe the hole dynamics induced by the attosecond XUV pulse. Due to the large bandwidth of the pulse, a manifold of ionization channels is open, thus leading to a superposition of many ionic states, i.e., to an electronic wave packet. For all open channels, the ionization amplitudes were quantitatively determined using the static-exchange Density Functional Theory method [5]. The evolution of the electronic wave packet has then been described by employing a standard time-dependent density matrix formalism. The results of the numerical simulations clearly show the production of an ultrafast electron dynamics characterized by oscillation frequencies in good agreement with the experimental observations [1].

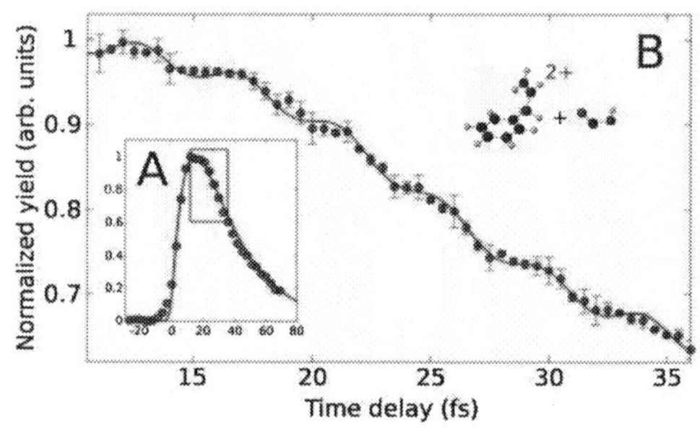

Figure 1. A) Fragmentation yield of doubly-charge immonium as function of the pump-probe delay. B) The same, measured with a 0.5-fs time step. The red line is the sum of the fitting curve shown in (A) and a sinusoidal function of period 4.3 fs.

References

[1] F. Calegari *et al*, 2014 *Science* **346** 6207 336
[2] J. Winkler *et al*, 2005 *Wiley-VCH* pp. 15-33
[3] L.S. Cederbaum *et al*, 1999 *Chem. Phys. Lett.* **307** 205
[4] F. Remacle *et al*, 2006 *P.N.A.S.* **103** 6793
[5] M. Stener *et al*, 2002 *Chem. Phys. Lett.* **351** 469

ph-O6

Time delay anisotropy in photoelectron emission from the isotropic ground state of helium

S. Heuser[1], A. Jiménez-Galán[2 1], C. Cirelli[1], M. Sabbar[1], R. Boge[1], M. Lucchini[1], L. Gallmann[1,4], I. Ivanov[5,6], A. Kheifets[5], J. M. Dahlström[7,8,9], E. Lindroth[7], L. Argenti[2], F. Martín[2,3,10], U. Keller[1]

[1] Physics Department, ETH Zurich, 8093 Zurich, Switzerland
[2] Departamento de Química, Módulo 13, Universidad Autónoma de Madrid, 28049 Madrid, Spain
[3] Instituto Madrileño de Estudios Avanzados en Nanociencia, Cantoblanco, 28049 Madrid, Spain
[4] Institute of Applied Physics, University of Bern, 3012 Bern, Switzerland
[5] Research School of Physics and Engineering,Australian National University, Canberra ACT 0200, Australia
[6] Center for Relativistic Laser Science, Institute for Basic Science (IBS), Gwangju 500-712, Rep. Korea, Asia
[7] Department of Physics, Stockholm University, AlbaNova University Center, SE-10691 Stockholm, Sweden
[8] Max Planck Institute for the Physics of Complex Systems, Noethnitzerstr. 38, 01187 Dresden, Germany
[9] Center for Free-Electron Laser Science, Luruper Chaussee 149, 22761 Hamburg
[10] Condensed Matter Physics Center (IFIMAC), Universidad Autónoma de Madrid, 28049 Madrid, Spain

Synopsis We show that time delays between electrons liberated from the $1s^2$ spherically symmetric ground state of helium depend on the emission direction of the electrons with respect to the polarization axis of the ionizing XUV light. This is attributed to the interplay between different final quantum states, which become accessible once two photons are involved in the photoionization process. This is a universal effect, which needs to be taken into account for any study dealing with photoionization dynamics.

The advent of attosecond science paved the way towards studying and understanding the nature of electron dynamics in atomic, molecular and condensed matter systems on their natural timescale. In particular, small delays in electron emission induced by single photon absorption have been measured with two different techniques: attosecond streaking and RABBITT[1]. Both methods are based on a pump-probe scheme, in which one pulse initiates the ionization process and the second pulse is used as a probe to retrieve the temporal information. So far, electron photoemission time delays have been measured between two different states of the same or different atoms. The dependence of the time delay on the electron emission direction, however, has not yet been investigated experimentally. When an electron is ionized by a single photon from an (n, ℓ) state, the dipole selection rules imply that the final photoelectron state has to be a $(E_f, \ell \pm 1)$ state. In the case of helium, starting from the $1s^2$ ground state, only one final state is available with a single photon absorption. However, if two photons are involved in the ionization process, as is the case in pump-probe schemes, two different final states are populated. The interplay between these final states may give rise to anisotropic time delays.
In this work, we report the first joint experimental and theoretical evidence of angle-dependent photoemission delays [2].Using the RABBITT technique, we measure a significant angular variation of the photoionization time delay of electrons ejected from the spherically symmetric $^1S^e$ ground state of helium, which can be as large as 50 attoseconds (Fig.1) for electrons emitted at an angle of 60 degrees with respect to the XUV polarization direction. The result is supported by virtually exact *ab initio* calculations which quantitatively reproduce the measured delays. We attribute the observed anisotropy to the interplay between the different quantum states that become accessible due to the two-photon transition process.
Our observation gives insight into a new general aspect of attosecond measurements of electron dynamics. The universal effect studied in this work is potentially relevant for the interpretation of angular integrated experiments and has to be taken into account in any study dealing with electron dynamics including more complex systems such as e.g. molecules and condensed matter.

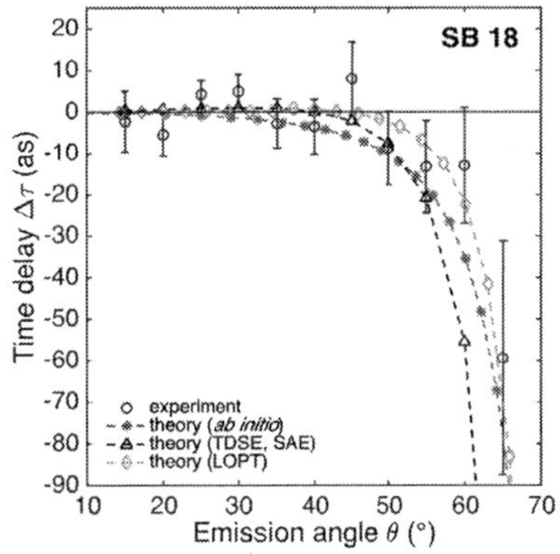

Figure 1. Angle-dependent photoemission time delay for photoelectrons with a kinetic energy of about 4.75eV.

References

[1] P.M. Paul *et al* 2001 *Science* **292** 1689

[2] S. Heuser *et al* 2015 arXive:1503.08966

[1]E-mail: alvaro.jimenez@uam.es

ph-O7

Optical oscillator strengths of the valence-shell excitations of molecular nitrogen measured by the dipole(γ,γ) method

Ya-Wei Liu*†, Xu Kang*†, Long-Quan Xu*†, Dong-Dong Ni*†, Xiao-Xun Mei*†, Lin-Fan Zhu*†[1]

* Hefei National Laboratory for Physical Sciences at Microscale and Department of Modern Physics, University of Science and Technology of China, Hefei, Anhui 230026, People's Republic of China

† Synergetic Innovation Center of Quantum Information and Quantum Physics, University of Science and Technology of China, Hefei, Anhui 230026, People's Republic of China

Synopsis Using the dipole(γ,γ) method, the optical oscillator strengths (OOSs) of the valence-shell excitations of molecular nitrogen were determined at a high energy-resolution of 70 meV and a high incident photon energy of about 10 keV. The present results are compared with the previous experimental and theoretical results.

As the most abundant molecule in the earth's atmosphere, photoionization of nitrogen resulting from its interaction with solar UV radiation plays an important role in the energy balance of the earth's upper atmosphere. In addition, the OOSs of nitrogen are also widely applied to plasma physics, laser, etc.

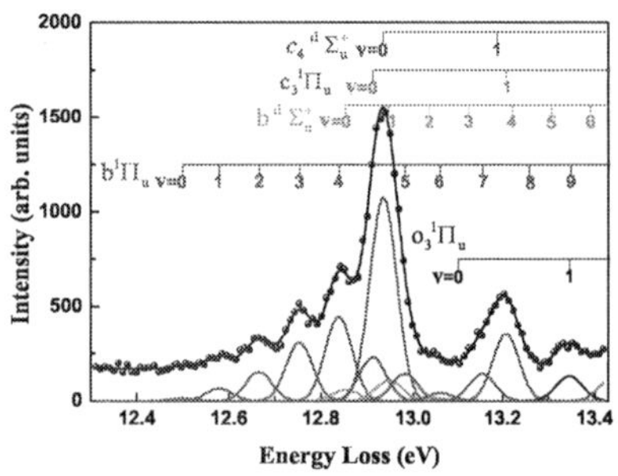

Figure 1. The IXS spectrum of molecular nitrogen at 2°.

With such importance and application value, the OOSs of the valence-shell excitations of molecular nitrogen were determined by the inelastic x-ray scattering (IXS) method in this work, which is operated at the negligibly small momentum transfer and is called the the dipole(γ,γ) method. In the measurement, the analyzer energy for the scattered photon was fixed at 9889.90 eV and the energy resolution was 70 meV. The IXS spectrum of molecular nitrogen at 2° is shown in Fig. 1 along with the vibrational states assigned. The present OOSs and the experimental errors are shown in Fig. 2.

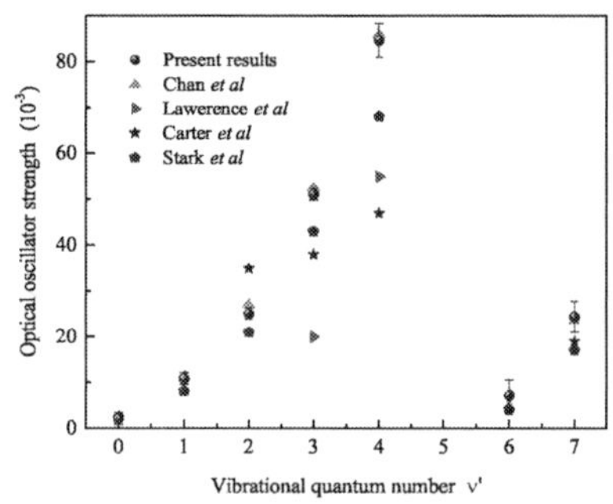

Figure 2. Absolute OOSs for $b^1\Pi_u$ of molecular nitrogen. The OOS of v'=4 of $b^1\Pi_u$ includes the contributions from nearby transitions due to the finite energy resolution.

It is obvious from Fig. 2 that the present results are agreement with the results of Chan *et al.* [1] measured by the dipole(e,e) method. However, our results are higher than the ones of Lawerence *et al.* [2], Carter *et al.* [3] and Stark *et al.* [4] measured by the photoabsorption method, which may be due to the line saturation effect. In summary, the present results provide an independent crosscheck for the measured OOSs of molecular nitrogen.

References

[1] W. F. Chan *et al* 1993 *Chem. Phys.* **170** 81

[2] G. M. Lawerence *et al* 1968 *J. Chem. Phys.* **48** 1989

[3] V. L. Carter *et al* 1972 *J. Chem. Phys.* **56** 4195

[4] G. Stark *et al* 2005 *J. Chem. Phys.* **123** 214303

[1]E-mail: lfzhu@ustc.edu.cn

ph-O8

Squared form factors for the $A^1\Pi$ and $B^1\Sigma^+$ vibronic bands of carbon monoxide studied by high-resolution inelastic x-ray scattering

Lin-Fan Zhu[1,2,†1], **Dong-Dong Ni**[1,2], **Xu Kang**[1,2], **Ya-Wei Liu**[1,2]

1 Hefei National Laboratory for Physical Sciences at Microscale and Department of Modern Physics, University of Science and Technology of China, Hefei, Anhui 230026, People's Republic of China

2 Synergetic Innovation Center of Quantum Information and Quantum Physics, University of Science and Technology of China, Hefei, Anhui 230026, People's Republic of China

Synopsis Using the inelastic x-ray scattering with a high resolution of 70 meV, squared form factors of molecular carbon monoxide have been determined from the ground state $X^1\Sigma^+$ to the vibronic states of $A^1\Pi$ and $B^1\Sigma^+$. Since the first Born approximation is satisfied in inelastic x-ray scattering, the inelastic squared form factors of the valence-shell excitations of carbon monoxide measured by this work not only provide the experimental benchmark data, but also serve as the high-energy limit for the electron impact method. Based on the present experimental results, the validity conditions of the previous electron impact works are discussed.

Carbon monoxide is the second most abundant interstellar molecule, which plays a crucial role in understanding the conditions and chemistry in the interstellar medium. Moreover, accurate and reliable spectroscopic data of carbon monoxide will make the interpretation of new astrophysical observations available. Consequently, the energy level structures and dynamic parameters of the valence-shell excitations of carbon monoxide have attracted great attention.

Figure 1 shows the present inelastic squared form factors (ISFF) for the excitation to $A^1\Pi(\nu' = 2)$ along with EELS values measured at different incident electron energies. It is clear that the EELS results at impact energies of 300 eV [1], 400 eV [1] and 500 eV [1] are in good agreement with the present IXS one, but at impact energies of 100 eV [2] and 200 eV [2], the EELS values deviate from the present IXS results with increasing momentum transfer. This phenomenon means that the FBA is satisfied at impact electron energies of more than 300 eV in the measured q^2 region, but does not hold for impact electron energies below 200 eV in the region $q^2 > 0.1$ a.u.. Excellent agreement is observed between the present results and the data by Zhong *et al.* [3] at an impact electron energy of 1500 eV in the region $q^2 < 0.7$ a.u.. However, the results of Zhong *et al.* disagree with the current measurement around the maximum,which may result from the inadequacy of the FBA when q^2 is beyond 0.7 a.u. even at the impact electron energy of 1500 eV.

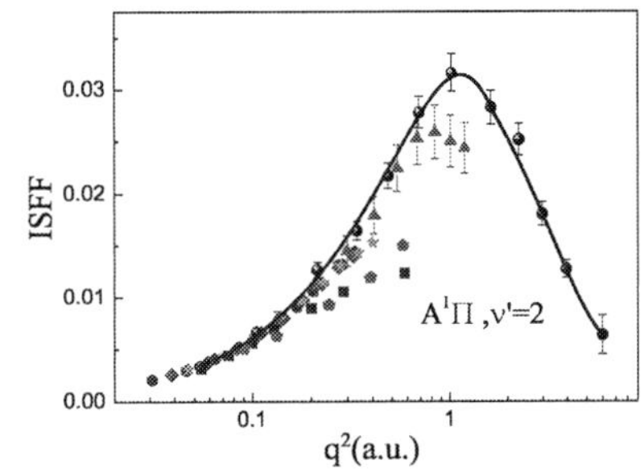

Figure 1. The dots are the present IXS results. The EELS results are: squares 100eV [2], pentagons 200eV [2], hexagon 300eV [1], diamonds 400eV [1], stars 500eV [1], and triangles 1500eV [3]

References

[1] A.Skerbele *at al* 1971 *J. Chem. Phys.* **55** 424

[2] H.Kato *at al* 2007 *J. Chem. Phys.* **126** 064307

[3] Zhong *at al* 1997 *Phys. Rev. A.* **55** 1799

[1]E-mail: lfzhu@ustc.edu.cn

ph-O9

Modulation of Attosecond Beating in Resonant Two-Photon Ionization

L. Argenti*[1], Á. Jiménez-Galán*, F. Martín*,†

* Departamento de Química, Módulo 13,Universidad Autónoma de Madrid, Cantoblanco 28049, Madrid, Spain

† Instituto Madrileño de Estudios Avanzados en Nanociencia (IMDEA-Nanociencia), Cantoblanco, 28049 Madrid, Spain

Synopsis We present a theoretical study of the photoelectron attosecond beating due to interference of two-photon transitions in the presence of autoionizing states. We show that, as a harmonic traverses a resonance, both the phase and frequency of the sideband beating significantly vary with photon energy. Furthermore, the beating between two resonant paths persists even when the pump and probe pulses do not overlap, thus permitting to reconstruct nonholographically coherent metastable wave packets. We characterize these phenomena with an analytical model that accounts for the effect of both intermediate and final resonances on two-photon processes.

Autoionizing states (AI) are hallmarks of electronic correlation, which shapes the reactivity of all many-body systems. Attosecond light sources provide a new means to investigate the role of such correlated states in atomic transitions, since one can map their evolution on a time scale smaller than their lifetime. The technique of reconstruction of attosecond beating in two-photon transitions (RABITT) [1], which has already been succesfully employed to study resonant processes [2], is particularly indicated to this task. So far, however, only transitions through either electronic bound states or autoionizing vibronic states without any appreciable contribution from the intermediate continuum have been considered. The phase of the complex transition amplitude was thus compatible with a change of π at the resonance energy.

In general, however, both the localized and continuum components of an intermediate AI contribute to the two-photon amplitude. We found that, when the two contributions are comparable, the phase exhibits a peaked excursion which returns to its original value, instead of a π jump. Furthermore, we predict that, for finite pulses, the frequency of the RABITT beating itself is altered. Models normally used to interpret these experiments do not contemplate the presence of resonances, while full *ab initio* calculations are either extremely demanding or outright unfeasible. For this reason, we developed a resonant analytical model which allows us to compute two-photon ionization spectra of attosecond pump-probe interferometric experiments, for arbitrary pulse sequences[3]. The model parameters can be retrieved either from experimental data or from selected *ab initio* simulations. We used this procedure to investigate, with RABITT, the two-photon ionization of helium in the region of the doubly excited states of the atom, obtaining excellent agreement with virtually exact *ab initio* simulations [4]. In particular, we show that both the frequency modulation and phase excursion of the transition through sp_2^+ and sp_3^+ ($^1P^o$) to $2p^2$ ($^1S^e$) are measurable.

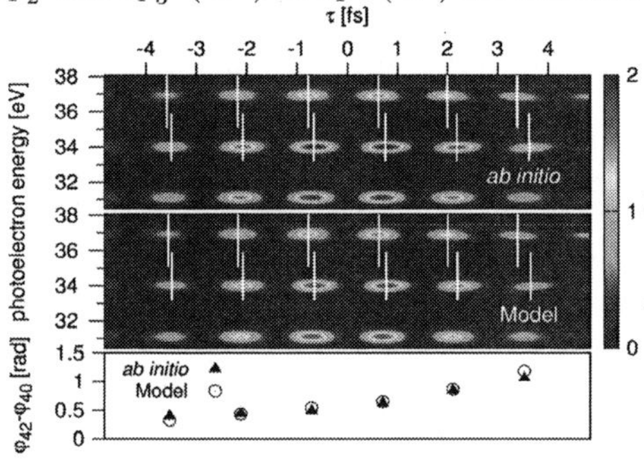

Figure 1. Photoelectron spectrum of He as a function of the pump-probe time delay. The presence of sp_2^+($\sim$ 35.5 eV) shifts the sidebands in opposite directions and modulates the beating frequency.

Finally, the beating between the overlapping sidebands of the two intermediate resonances persist even when the pump and probe pulses are separated. From this beating it is possible to reconstruct the metastable wavepacket generated by the pump pulse.

References

[1] P. M. Paul *et al.* 2001 *Science* **292**, 1689

[2] M.Swoboda *et al.* 2010 *Phys.Rev.Lett.* **104**, 103003

[3] Á. Jiménez-Galán, L. Argenti and F.Martín 2014 *Phys.Rev.Lett.* **113**, 263001

[4] L. Argenti and E. Lindroth 2010 *Phys. Rev. Lett.* **105**, 053002

[1]E-mail: luca.argenti@uam.es

The $1s^{-1}2s^{-1}$ and $1s^{-1}2p^{-1}$ double core-hole shake-up satellites in Argon

R. Püttner[*,1], G. Goldsztejn[†,#], D. Céolin[§], J.-P. Rueff[§], T. Moreno[§], K. Kushawaha[†,#], T. Marchenko[†,#], R. Guillemin[†,#], R. L. Journel[†,#], M. N. Piancastelli[†,#,‡], and M. Simon[†,#,§]

[*]Institut für Experimentalphysik, Freie Universität Berlin, Arnimallee 14, D-14195 Berlin, Germany
[†]CNRS, UMR 7614, Laboratoire de Chimie Physique-Matière et Rayonnement, F-75005, Paris, France
[#] Sorbonne Universités, UPMC Univ Paris 06, UMR 7614, LCPMR, F-75005, Paris, France
[§]Synchrotron SOLEIL, l'Orme des Merisiers, Saint-Aubin, FR-91192 Gif-sur-Yvette Cedex, France
[‡]Department of Physics and Astronomy, Uppsala University, SE-75120 Uppsala, Sweden

Synopsis The Ar $1s^{-1}2s^{-1}$Ryd and the $1s^{-1}2p^{-1}$Ryd double core-hole shake-up satellites are measured and assigned. From these assignments the binding energies of the doubly core-ionized states are derived. The Ar $2p^{-1}$ core hole reduces the Ar $1s^{-1}$ lifetime broadening significantly. The observation is rather general and has also been reported for a number of Si, S, and Cl containing molecules.

Using a new experimental setup for high-resolution HAXPES (HArd X-ray PhotoElectron Spectroscopy) in gas phase, which is mounted at the GALAXIES beamline of SOLEIL [1], we investigated the $1s^{-1}2s^{-1}$ and the $1s^{-1}2p^{-1}$ shake up structures of Argon, i.e. an Ar 1s photoionization accompanied by a shake of a 2s or 2p inner shell electron into an unoccupied Rydberg orbital [2]. The spectrum of the $1s^{-1}2p^{-1}$ shake up satellites is dominated by the np (n = 4 to 7) Rydberg series converging towards the ionization thresholds $^1P_1$ and $^3P_{2,1,0}$, which are determined with high accuracy. These values agree reasonably well with theoretical predictions by Kuetgens and Hormes [3] but differ significantly from previous experimental values [4].

For the $1s^{-1}2p^{-1}$np Rydberg states a lifetime broadening of 675(20) meV was derived from the fit analysis. This value is considerably smaller than the sum of the lifetime broadenings of the Ar 1s (655 meV [5]) and the Ar 2p (115 meV [6]) state of 770 meV. This observation can be explained with a significantly smaller Ar 1s lifetime broadening caused by the presence of a Ar 2p core hole, in full agreement with theory [7].

The $1s^{-1}2p^{-1}$ shake up structures match well the Ar $1s^{-1}$ photoabsorption spectrum. However, they exhibit more details due to a larger splitting between the resonances at a similiar lifetime broadening. This provides new information about the spectroscopy of both the $1s^{-1}2p^{-1}$ state as well as the photoabsorption spectrum of the neutral atom.

Below the $1s^{-1}2s^{-1}$ thresholds the spectrum is dominated by the broad $1s^{-1}2s^{-1}$ (^{1,3}S) 4s Rydberg states which are split by approximately 16 eV. Moreover, a number of narrow lines are observed above the $1s^{-1}2p^{-1}$ (^{1,3}P) thresholds. Based on linewidths and splittings similar to those of the $1s^{-1}2p^{-1}$ (^{1,3}P) np Rydberg states these narrow lines are assigned to $1s^{-1}2p^{-1}$ (^{1,3}P) $3p^{-1}$n'l'n''l'' final states populated via double shake processes.

In addition, the $1s^{-1}2p^{-1}$ shake up satellites are also observed in molecules for a number of Si, S, and Cl containing molecules showing that this is a rather general effect. This observation opens the possibility to study single-site double core-hole (SS DCH) states using conventional photoelectron spectroscopy at synchrotron radiation facilities.

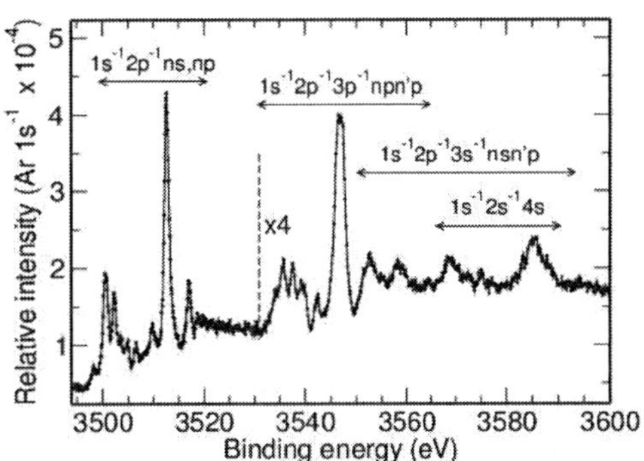

Figure 1: The Ar $1s^{-1}2s^{-1}$Ryd and the $1s^{-1}2p^{-1}$Ryd double core-hole shake-up satellites

References

[1] J. P. Rueff *et al.* 2015 J. Synchrotron Rad. **22** 175
[2] R. Püttner *et al.* 2015 Phys. Rev. Lett. **114** 193001
[2] U. Kuetgens and J. Hormes 1991 Phys. Rev. A **44** 264
[3] P.-A. Raboud *et al.* 2002 Phys. Rev. A **65** 062503
[4] D. Céolin *et al.* 2013 J. Electron Spectrosc. Relat. Phenom. **190 188**
[5] K. C. Prince *et al.* 1999 J. Electron Spectrosc. Relat. Phenom. **101-103** 141
[6] C. P. Bhalla 1973 Phys. Rev. A **8** 2877

[1]E-mail: puettner@physik.fu-berlin.de

ph-O11

Keldysh nonequilibrium Green's function vs. Feshbach projection operator approach for plasmon-assisted photoemission

Yaroslav Pavlyukh [1], **Michael Schüler, Jamal Berakdar**

Institut für Physik, Martin-Luther-Universität Halle-Wittenberg, 06120 Halle, Germany

Synopsis A unified theoretical treatment of the single and double electron emission is achieved by using the Feshbach projection method. In this formalism the final target's state fixes the projection operator which subsequently determines the effective Hamiltonian and the optical potential for emitted electrons. The method of non-equilibrium Green's functions is a complementary approach which also allows to treat such processes diagrammatically. We explicitly establish a correspondence between these two approaches and illustrate the diagrammatic technique by calculations of the two-electron emission from C_{60} assisted by the excitation of plasmons.

Starting from the explicit form of the projection operators dividing the whole Hilbert space of the system into that of the emitted electron(s) and the target we derive the effective one- and two-particle Hamiltonian, discuss integral equations for the Green's functions describing emitted particles and demonstrate a close connection of this formalism to the nonequilibrium Green's function theory. For the latter, one derives the diagrammatic expansions for one- and two-particle currents starting from the time-dependent perturbation theory and using the adiabatic switching of the electron-electron interaction.

The diagrammatic structure of one- and two-particle currents is surprisingly simple: one starts with the density-density response function $\chi^<$ which necessarily contains two blocks associated with the forward ("−") and backward ("+") parts of the Keldysh contour. Requesting that one or two lines flowing from "−" to "+" blocks are associated with scattering states (with momenta $\vec{k}_i$) one obtains exactly the diagrams for SPE and DPE currents showing the close connection between these types of light-matter interaction (Fig. 1).

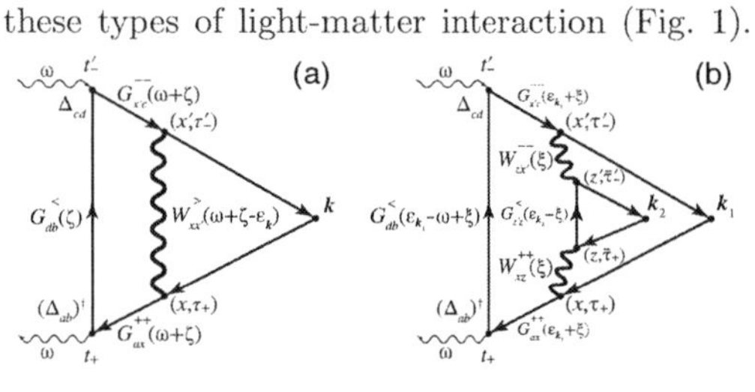

Figure 1. (a) SPE diagram with external plasmonic losses, (b) DPE diagram describing a related plasmon assisted process.

Finally, we present a detailed analysis of the plasmon-assisted DPE and show that if one of the emitted particles is unobserved, its diagrammatic representation reduces to the one describing external losses in the SPE process considered by Caroli *et al.* [2]. Plasmon pole approximation is employed to derive computationally manageable expressions and to perform calculations the C_{60} molecule [1].

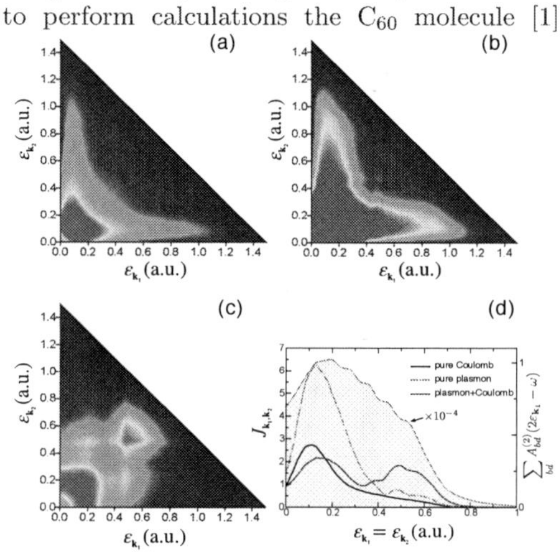

Figure 2. Two-electron current as a function of the photoelectron energies for incident photon energy $\omega = 2.0$ a.u. and plasmon energy $\omega_p = 0.8$ a.u. (a) The process is mediated by the pure Coulomb interaction. (b) Plasmonic contribution. (c) Total signal including the interference terms. (d) Equal energy sharing ($\varepsilon_{\vec{k}_1} = \varepsilon_{\vec{k}_2}$) for the current and trace of the two-particle spectral density (shaded curve).

[1] E-mail: yaroslav.pavlyukh@physik.uni-halle.de

References

[1] Y. Pavlyukh, M. Schüler and J. Berakdar 2015 *Phys. Rev. B* **91** 155116

[2] C. Caroli, D. Lederer-Rozenblatt, B. Roulet, and D. Saint-James 1973 *Phys. Rev. B* **8** 4552

ph-O12

Interatomic Coulombic Decay Processes after Multiple Valence Excitations in Ne Clusters

D. Iablonskyi[a 1], K. Nagaya[b], H. Fukuzawa[a] K. Motomura[a], Y. Kumagai[a], S. Mondal[a], T. Tachibana[a], T. Takanashi[a], T. Nishiyama[b], K. Matsunami[b], P. Johnsson[c], P. Piseri[d], G. Sansone[e], A. Dubrouil[e], M. Reduzzi[e], P. Carpeggiani[e], C. Vozzi[e], M. Devetta[e], M. Negro[e], D. Faccialà[e], F. Calegari[e], A. Trabattoni[e], M. Castrovilli[e], Y. Ovcharenko[f], T. Möller[f], M. Mudrich[g], F. Stienkemeier[g], M. Coreno[h], M. Alagia[i], B. Schütte[j], N. Berrah[k], C. Callegari[l], O. Plekan[l], P. Finetti[l], C. Spezzani[l], E. Ferrari[l], E. Allaria[l], G. Penco[l], C. Serpico[l], G. De Ninno[l], B. Diviacco[l], S. Di Mitri[l], L. Giannessi[l], K. C. Prince[i,l], M. Yao[b], and K. Ueda[a]

[a] Institute of Multidisciplinary Research for Advanced Materials, Tohoku University, 980-8577 Sendai, Japan
[b] Department of Physics, Graduate School of Science, Kyoto University, 606-8502 Kyoto, Japan
[c] Department of Physics, Lund University, P. O. Box 118, 22100 Lund, Sweden
[d] Dipartimento di Fisica, Università degli Studi di Milano, 20133 Milano, Italy
[e] CNR-IFN, Piazza Leonardo da Vinci 32, 20133 Milan, Italy
[f] Institut für Optik und Atomare Physik, TU Berlin, 10623 Berlin, Germany
[g] Physikalisches Institut, Universität Freiburg, 79104 Freiburg, Germany
[h] CNR-ISM, Area della Ricerca di Roma 1, Monterotondo Scalo, Italy
[i] CNR-IOM, Area Science Park, 34149 Trieste, Italy
[j] Max-Born-Institut, Max-Born-Strasse 2 A, 12489 Berlin, Germany
[k] Department of Physics, University of Connecticut, Storrs, CT 06269, USA
[l] Elettra-Sincrotrone Trieste, Area Science Park, 34149 Trieste, Italy

Synopsis We present a comprehensive analysis of autoionization processes in Ne clusters (~5000 atoms) after multiple valence excitations by free electron laser radiation. The evolution from 2-body interatomic Coulombic decay (ICD) to 3-body ICD is demonstrated when changing from surface to bulk Frenkel exciton excitation. Super Coster-Kronig type 2-body ICD is observed at Wannier exciton which quenches the main ICD channel.

Previously, Yase *at al* [1] investigated multiple excitation of Wannier type excitons (corresponding to the $2p \to 3d$ atomic resonance) at 20.26 eV in Ne clusters by the intense extreme ultraviolet free electron laser (EUV-FEL) at SCSS (SPring-8 Compact SASE Source, Japan) and found that the electron emission is dominated by low energy electron emission that originates from a nanoplasma.

In the present experiment, we have extended our observations to multiple excitations of surface and bulk Frenkel type excitons (corresponding to the $2p \to 3s$ atomic resonance) at 17.12 eV and 17.65 eV, respectively, using the new seeded EUV-FEL, FERMI (Trieste, Italy) [2].

At the lowest surface Frenkel exciton we can clearly see the pure 2-body ICD peak at ~11.5 eV, predicted by Kuleff *at al* [3], with its multistep ICD tail which is similar to direct multistep ionization in Ar clusters [4]. The situation changes for the bulk Frenkel exciton, where the broad structure around 5 eV is identified as 3-body ICD of knock-off type (also known as collective autoionization [5]) and becomes dominant over 2-body ICD at high FEL intensities. For the Wannier exciton we can see complete quenching of the main 2-body ICD by super Coster-Kronig type ICD in which one 3d electron relaxes to a 3s orbital and another 3d electron is ejected with ~1.8 eV kinetic energy (see Fig. 1).

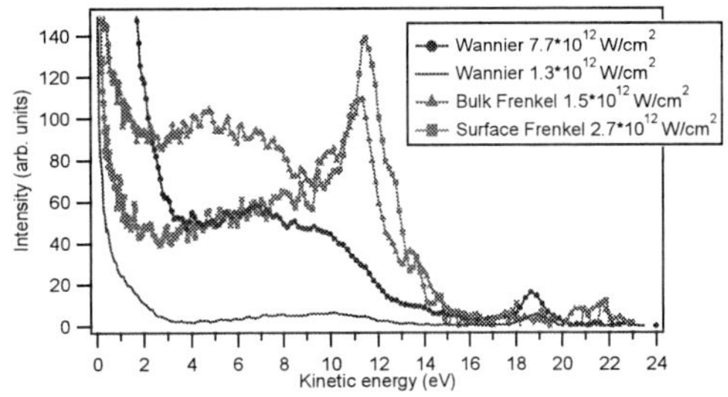

Figure 1. Electron emission spectra for excitation of different excitons and selected FEL intensities.

References

[1] S. Yase *et al.* 2013 *Phys. Rev. A* **88** 043203

[2] E. Allaria *et al.* 2012 *Nat. Photonics* **6** 699

[3] A. I. Kuleff *et al.* 2010 *Phys. Rev. Lett.* **105** 043004

[4] C. Bostedt *et al.* 2008 *Phys. Rev. Lett.* **100** 133401

[5] Y. Ovcharenko *et al.* 2014 *Phys. Rev. Lett.* **112** 073401

[1]E-mail: denys@tagen.tohoku.ac.jp

POSTER CONTRIBUTIONS

Photoelectron angular distributions and correlations in sequential two-photon double ionization by circularly polarized XUV radiation

Alexei N. Grum-Grzhimailo* [1], Elena V. Gryzlova*, Ekaterina I. Staroselskaya†

* Skobeltsyn Institute of Nuclear Physics, Lomonosov Moscow State University, Leninskie gory, 119991, Moscow, Russian Federation

† Physical Department, Lomonosov Moscow State University, Leninskie gory, 119991, Moscow, Russian Federation

Synopsis Photoelectron angular distributions (PADs) and angular correlation functions in sequential two-photon double ionization of atoms by circularly polarized radiation are discussed taking into account first-order corrections to the dipole approximation caused by the interference of electric dipole and quadrupole amplitudes. The non-dipole contributions violate the symmetry of PADs and angular correlation functions already at photon energies of about 50 eV.

The development of free-electron laser (FEL) facilities providing intense radiation in the XUV range has stimulated numerous investigations of nonlinear atomic processes. One of the simplest examples is sequential two-photon double ionization (2PDI) which may be considered as the following two-step process: (a) ionization of the neutral atom and (b) ionization of the ion by a second photon from the same pulse with possible evolution of the intermediate ion in the time interval between the absorption of the first and the second photon. The sequential 2PDI of the noble gases,

a) $A(np^6\,{}^1S) + \gamma \rightarrow A^+(np^5\,{}^2P_{3/2,1/2}) + e_1$,

b) $A^+(np^5\,{}^2P_{3/2,1/2}) + \gamma \rightarrow A^{++}(np^4\,{}^3P, {}^1D, {}^1S) + e_2$,

by linearly polarized radiation was studied both experimentally and theoretically, including pioneering measurements of PADs by the group of Uwe Becker [1]. The angular correlation functions have been observed in [2]. Nowadays experimentalists can also use circularly polarized XUV radiation generated by the FEL FERMI. In contrast to the case of linearly polarized field, the PAD for the circularly polarized radiation keeps axial symmetry with respect to the photon beam regardless to the radiation field multipoles involved. Therefore we expect that using circular radiation might be more convenient in experiments with VMI spectrometers and hence better clarify the dynamics of the process.

To treat theoretically the sequential 2PDI by circularly polarized radiation we apply an approach developed in [3,4]. Figure 1 shows angular correlation functions in the 2PDI of Ar to the residual ionic $Ar^{++}(3p^4\,{}^3P)$ state after incoherent summation over the $Ar^+(3p^5\,{}^2P_{3/2,1/2})$ fine structure states of the intermediate ion.

At higher photon energies the angular correlation pattern (right panel) is similar for Ne, Ar, and Kr. In particular, the pattern is symmetric with respect to the replacement $\theta_{1(2)} \rightarrow \pi - \theta_{1(2)}$. This symmetry is violated in Ar at lower photon energies (left panel) due to the increase of the relative non-dipole contribution closer to the Cooper minimum in the 3p→εd partial cross section.

The results for other terms of the residual ion, for other atoms (neon and krypton), and other photon energies will be presented and analysed at the conference [5].

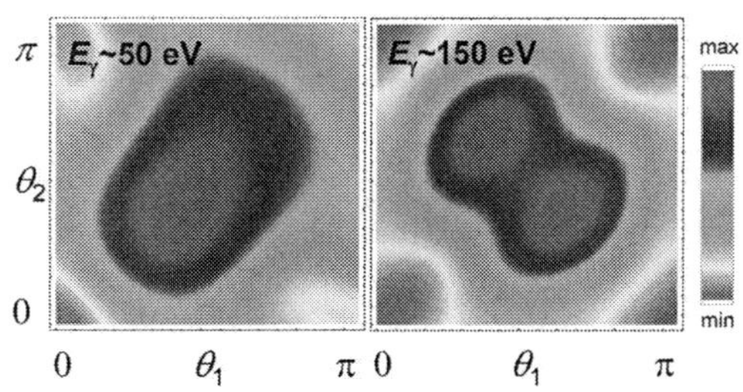

Figure 1. Angular correlation functions in sequential 2PDI of Ar (see text). Z-axis is directed along the photon beam and both photoelectrons are detected in the same plane $\varphi = 0$.

References

[1] M. Braune, A. Reinköster, J. Viefhaus, B. Lohmann, and U. Becker 2007 ICPEAC, Freiburg, Germany, Book of Abstracts, Fr034;
M. Braune *et al.* 2015 J. Mod. Opt. (accepted)

[2] M. Kurka *et al* 2009 J. Phys. B **42** 141002

[3] S. Fritzsche *et al* 2008 J. Phys. B **41** 165601

[4] A.N. Grum-Grzhimailo *et al* 2012 J. Phys. B **45** 215602

[5] E.V. Gryzlova *et al* 2014 J. Phys. B **47** 195601;
A.N. Grum-Grzhimailo *et al* 2015 J. Mod. Opt. (accepted)

[1] E-mail: algrgr1492@yahoo.com

P2

Theory of ultrafast x-ray photoelectron diffraction

Shota Tsuru[1], Kyo Nakajima[2], Tokuei Sako[3], Takashi Fujikawa[1] and Akira Yagishita[2]

[1]Graduate School of Advanced Integration Science, Chiba University, Chiba 263-8522, Japan
[2]Institute of Materials Structure Science, KEK, Tsukuba, Ibaraki 305-0801 Japan
[3]Laboratory of Physics, CST, Nihon University, 7-24-1, Narashinodai, Funabashi, 274-8501 Japan

Synopsis Ultrafast x-ray photoelectron diffraction can realize molecular movies with femtosecond time and Ångstrom spatial resolution for small- to medium-sized gas-phase molecules. Here we theoretically demonstrate the diffraction images of simple molecules during the photochemical reactions.

A pump and prove experiment of ultrafast optical laser and x-ray free-electron laser (XFEL) pulses can realize ultrafast x-ray photoelectron diffraction (XPD) measurements. The XPD patterns capture the entire molecular geometry, thus such experiment can track atomic positions during unimolecular photochemical reactions in femtosecond time scale and Ångstrom spatial resolution. Hence, we theoretically demonstrate the XPD images of simple molecules during photochemical reactions. Under the assumption that a valence electron in the ground state, $\Phi_e(\boldsymbol{r};\boldsymbol{R})\chi_v^e(\boldsymbol{R})$, is excited to an excited state, $\Phi_g(\boldsymbol{r};\boldsymbol{R})\chi_0^g(\boldsymbol{R})$, we solve time-dependent Schrödinger equation of such excitation process within the Born-Oppenheimer approximation. With the use of the transition dipole moment function of $\boldsymbol{d}_{eg}(\boldsymbol{R})=\langle\Phi_e(\boldsymbol{r};\boldsymbol{R})|\boldsymbol{r}|\Phi_g(\boldsymbol{r};\boldsymbol{R})\rangle$ within the first-order perturbation approximation the vibrational wave packet in the excited state is written as

$$X_e(\boldsymbol{R},\mathrm{t})=-i\sum_v\langle\chi_v^e(\boldsymbol{R})|\hat{e}\cdot\boldsymbol{d}_{eg}(\boldsymbol{R})|\chi_0^g(\boldsymbol{R})\rangle \times\chi_v^e(\boldsymbol{R})e^{-iE_v^e t}I(t;\omega,\omega_{v0})$$

,where $I(t;\omega,\omega_{v0})=\int_0^t d\tau E^{OL}(\tau)\cos\omega\tau e^{-i\omega_{v0}\tau}$. $E^{OL}(\tau)$, ω, and $\hat{e}$ are the amplitude, frequency, and polarization vector of the optical laser, respectively. ω_{v0} is the energy difference between the ground and excited states ; $\omega_{v0}=E_v^e-E_0^g$. Then, we calculate the XPD images at the instant probed by the XFEL pulses;

$$\frac{d\sigma}{d\hat{\boldsymbol{k}}}(t)\propto\int d\boldsymbol{R}\left|\sum_v\langle\chi_v^e(\boldsymbol{R})|\hat{\boldsymbol{e}}\cdot\boldsymbol{d}_{eg}(\boldsymbol{R})|\chi_0^g(\boldsymbol{R})\rangle \times\chi_v^e(\boldsymbol{R})e^{-iE_v^e t}I(t;\omega,\omega_{v0})\right|^2\left|\langle\phi_{\boldsymbol{k}}^-(\boldsymbol{r})|\boldsymbol{E}^{FEL}\cdot\boldsymbol{r}|\phi_c(\boldsymbol{r})\rangle_R\right|^2.$$

The XPD amplitude at the molecular geometry $\boldsymbol{R}$ is given by the multiple-scattering theory based on the sudden approximation for electronic transitions [1]. At the conference, we present a few calculational results of ultrafast photoelectron diffractions.

References

[1] Shinotsuka H *et al* 2008 *Phys. Rev.* B 77, 085404

Two-photon triple ionization of lithium

James Colgan* [1], **M. S. Pindzola**†

* Theoretical Division, Los Alamos National Laboratory, Los Alamos, NM 87545, USA
† Department of Physics, Auburn University, Auburn AL 36849, USA

Synopsis We use the time-dependent close-coupling approach to examine two-photon triple ionization of lithium at a photon energy of 115 eV. The resulting angular distributions for three outgoing electrons are computed and compared to similar distributions arising from single-photon triple ionization.

Triple ionization of atoms by photon absorption leads to the complex scenario of three electrons moving in a Coulomb field, a good example of the four-body Coulomb problem. If the outgoing energies of the electrons are not too large, the electron-electron interactions will dominate the resulting angular distributions of the ionized electrons. Such four-body Coulomb problems have been examined in the single-photon triple ionization of Li [1, 2, 3] at a variety of photon energies, and also in the electron-impact double ionization of helium [4, 5]. Several of these studies examined the low-energy break-up patterns and found that either a 'triangular' or 'T-shape' break-up pattern was most likely.

In this work, we report on the triple ionization of Li after absorption of two 115 eV photons using the time-dependent close-coupling approach. Two-photon double ionization of helium has been extensively studied by now many groups in recent years. One of the major findings of the helium studies was that the angular distribution of the two outgoing electrons is quite different for two-photon ionization compared to single-photon ionization. This is chiefly because of the strong selection rules that dominate the angular patterns for two electrons in the final (odd) angular momentum $L = 1$ state that is the result of single-photon absorption. In two-photon absorption the final states are either (even) $L = 0$ or $L = 2$, and the selection rules that govern these states result in quite different angular distributions. In the *triple* ionization case, we explore the angular distributions for three outgoing electrons after two-photon absorption. In contrast to the helium two-photon absorption case, we find that the angular distributions for two-photon absorption are quite similar to those for single-photon absorption. We explore the reasons for these similarities and also discuss the numerical techniques of our time-dependent approach to this problem.

The Los Alamos National Laboratory is operated by Los Alamos National Security, LLC for the National Nuclear Security Administration of the U.S. Department of Energy under Contract No. DE-AC5206NA25396.

References

[1] A. Emmanouilidou, P. Wang and J. M. Rost 2008 *Phys. Rev. Letts.* **100** 063002

[2] J. Colgan and M. S. Pindzola 2012 *Phys. Rev. Letts.* **112** 023202

[3] J. Colgan, A. Emmanouilidou, M. S. Pindzola 2013 *Phys. Rev. Letts.* **110** 063001

[4] M. Dürr et al 2007 *Phys. Rev. Letts.* **98** 193201

[5] X. Ren, A. Dorn, and J. Ullrich 2008 *Phys. Rev. Letts.* **101** 093201

[1] E-mail: jcolgan@lanl.gov

P4

Dressing Effects in the Attosecond Transient Absorption Spectra of Doubly-Excited States in Helium

L. Argenti*[1], Á. Jiménez-Galán*, C. Marante*, C. Ott†‡, T. Pfeifer†‡, F. Martín*¶§

* Departamento de Química, Módulo 13,Universidad Autónoma de Madrid, Cantoblanco 28049, Madrid, Spain
† Max-Planck Institut für Kernphysik, Saupfercheckweg 1, 69117 Heidelberg, Germany
‡ Center for Quantum Dynamics, Ruprecht-Karls-Universität Heidelberg, 69120 Heidelberg, Germany
§ Condensed Matter Physics Center (IFIMAC), Universidad Autónoma de Madrid, 28049 Madrid, Spain
¶ IMDEA-Nanociencia, Cantoblanco, 28049 Madrid, Spain

Synopsis Strong-field manipulation of autoionizing states is a crucial aspect of electronic quantum control. Recent measurements of the attosecond transient absorption spectrum (ATAS) of helium dressed by a few-cycle visible pulse [C. Ott *et al*, Nature **516**, 374 (2014)] provide evidence of the inversion of Fano profiles. With the support of accurate *ab-initio* calculations that reproduce the results of the latter experiment, here we investigate the new physics that arise from ATAS when the laser intensity is increased. In particular, we show that (i) previously unnoticed signatures of the dark $2p^2\ ^1$S doubly excited state are observed in the experimental spectrum, (ii) inversion of Fano profiles is predicted to be periodic in the laser intensity, and (iii) the ac-Stark shift of the higher terms in the $sp_{2,n}^+$ autoionizing series exceeds the ponderomotive energy, which is the result of a genuine two-electron contribution to the polarization of the excited atom.

ATAS [1] has affirmed itself as a valid technique to monitor and control electronic wavepackets [1–3]. Control is achieved with a laser pulse that imparts an ac-Stark shift to the wavepacket components, thus altering their relative phases. The latter map onto the asymmetry of their resonant profiles in the ATAS spectrum [4]. This principle has been recently applied to a two-electron wave packet in the helium atom, formed by the $sp_{2,n}^+\ ^1$P^o doubly excited states [3]. The experiment evidenced the Autler–Townes splitting of the $sp_{2,2}^+$ state, which is coupled to the $2p^2\ ^1$S state, as well as inversion of the Fano profile in the higher terms of the $sp_{2,n}^+$ series as a function of both the time delay between the XUV and the VIS pulses, and the intensity of the VIS dressing field [4].

Here we show, by means of accurate *ab-initio* calculations [3, 5, 6] that reproduce the experimental results for the helium atom, that ATAS can be used to measure the non-perturbative response of autoionizing states to external dressing fields. In particular, we show that the Autler-Townes splitting of the dressed $2p^2\ ^1$S "dark" state can be measured, and that the asymmetric resonant features in the spectrum undergo repeated inversions as the intensity of the driving laser is increased. The ac-Stark shift of the doubly-excited states, which are important parameters for the design of quantum-control protocols, increases non-linearly with the intensity of the dressing field; for the higher states of the autoionizing $sp_{2,n}^+$ series, it exceeds by more than twice the theoretical single-active-electron limit. Finally, we establish a relationship between the ATAS spectrum and the non-diagonal components of the XUV electrical susceptibility, thus extending multidimensional spectroscopy to the XUV domain (Fig. 1), which provides an alternative way to measure shifts of and coherences between states above the ionization threshold.

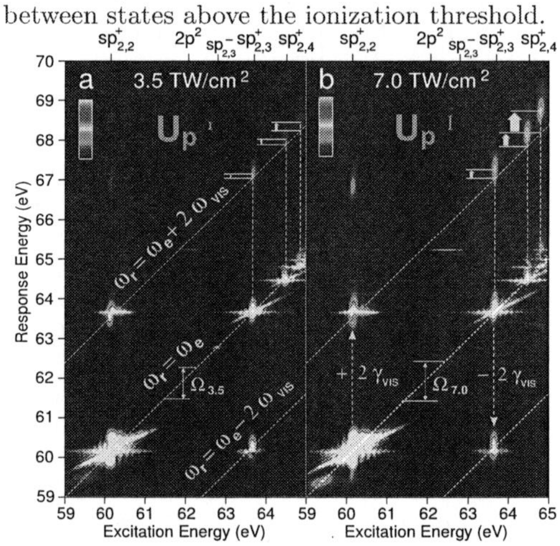

Figure 1. Reconstruction of the non-diagonal XUV electrical susceptibility $|\chi_{\rm nd}(\omega_r, \omega_e)|^2$ of dressed helium.

References

[1] E. Goulielmakis *et al.* 2010 *Nature* **466**, 739.
[2] A. R. Beck *et al.* 2014 *New J. Phys.* **16**, 113016.
[3] C. Ott *et al.* 2014 *Nature* **516**, 374.
[4] C. Ott *et al.* 2013 *Science* **516**, 374.
[5] L. Argenti and E. Lindroth 2010 *PRL* **105**, 053002.
[6] L. Argenti *et al.* 2015 *submitted.*

[1]E-mail: luca.argenti@uam.es

Excitation of vibrational modes in the ionization of water molecule by XUV/X-ray radiation

Selma Engin[† 1], **Jesús González-Vázquez**[†], **Inés Corral**[†], **Alicia Palacios**[†], **David Ayuso**[†], **Piero Decleva**[**], **and Fernando Martín**[†‡]

[†] Departamento de Química, Módulo 13, Universidad Autónoma de Madrid, Cantoblanco 28049 Madrid, Spain
[‡] Instituto Madrileño de Estudios Avanzados en Nanociencia, Cantoblanco 28049 Madrid, Spain
[*] Dipartimento di Scienze Chimiche e Faramaceutiche, Università di Trieste, 34127 Trieste, Italy
[*] CNR-IOM, Trieste, Italy

Synopsis We present a theoretical study of the vibrationally resolved core photoionization of the water molecule up to high photon energies. In order to understand the role of the coupled electron-nuclear motion in polyatomic molecules, we thus have implemented a new methodology to describe all vibrational modes of a polyatomic molecule. We show our preliminary results on the O(1s) photoionization, with special focus on the vibrationally resolved cross sections in a large range of photon energies, reaching up to 1500 eV.

Most theoretical studies of coupled electron and nuclear dynamics in photoionization problems have been restricted to diatomic molecules [1] or to a single vibrational mode of polyatomic molecules [2]. In some cases, a reasonable approximation is to reduce the nuclear degrees of freedom to one principal normal mode. This was shown in photoionization problems for highly symmetric targets such that CH_4 or BF_3 [2], whose symmetric stretching mode was the main active mode upon the photon absorption. However, such simplification is not valid in most problems.

The aim of this study is to extend the methodology previously developed in those molecular targets, which used a one-dimensional vibrational wave function, to include an arbitrary number of nuclear degrees of freedom. We thus seek to achieve a complete description of vibrational motion, although, for simplicity our first implementation uses Jacobi coordinates, which are particularly suitable for non-linear triatomic molecules. In the present study, we report results for the O(1s) photoionization of the water molecule induced by synchroton radiation, where all normal modes are simultaneously accounted for : bending, symmetric stretching and asymmetric stretching. Accurate vibrationally resolved cross sections require the computation of dipole matrix elements for the transition from the ground state of the neutral to the core-ionized water molecule. These dipole elements are obtained for the full grid of nuclear coordinates using the static-exchange DFT method developed by P. Decleva and collaborators [3]. The reliability of our approach is tested by comparing our calculated O(1s) vibrationally resolved photoelectron spectrum with available experimental data for a photon energy of 590 eV [4]. We find a good agreement with the vibrational progression at 590 eV, where the dominant contribution is that of the bending mode of the core-ionized water molecule. We further analyze the v-ratios as a function of photon energy, i.e. the ratios between the photoionization cross sections for individual vibrational states. For all normal modes, we observe oscillations around the Franck-Condon factor value, similar to those found in previous works for other diatomic and polyatomic targets (CO, CH_4, BF_3, etc) [1, 2], whose origin comes from diffraction of the ejected electron by the surrounding H atoms.

References

[1] E. Kukk *et al.*, *Phys. Rev. A* **88** 033412 (2013)
D. Ayuso *et al.*, *J. Elec. Spec. Rel. Phen.* **195** 320-326 (2014)

[2] E. Plésiat *et al.*, *Phys. Rev. A* **85** 023409 (2012)
K. Ueda *et al.*, *J. Chem. Phys.* **139** 124306 (2013)

[3] M. Stener *et al.*, *J. Chem. Phys.* **112** 10871 (2000)
M. Stener *et al.*, *Chem. Phys. Lett.* **351** 469 (2002)
D. Toffoli *et al.*, *Chem. Phys.* **276** 2543 (2002)

[4] R. Sankari *et al.*, *Chem. Phys. Lett.* **380** 647-653 (2003)

[1]E-mail: selma.engin@uam.es

P6

Photoionization time delays in molecular hydrogen.

R. Bello*[1], Sebastian Heuser†, A. Palacios*, Matteo Lucchini†, Lukas Gallmann†, Claudio Cirelli†, F. Martín*◇, Ursula Keller†

* Departamento de Química, M-13, Universidad Autónoma de Madrid, Cantoblanco 28049, Madrid, Spain
◇ Instituto Madrileño de Estudios Avanzados (IMDEA) en Nanociencia, Cantoblanco, 28049 Madrid, Spain
† Physics Department, ETH Zurich, 8093 Zurich, Switzerland

Synopsis We present our preliminary theoretical and experimental results on the use of RABBIT techniques to extract the photoionization time delays of H_2 in the autoionization region.

Laser assisted photoionization processes provide an elegant tool to control electron dynamics in atoms and molecules in real time [1]. A sucessfull strategy is the RABBIT technique (reconstruction of attosecond beating by interference of two-photon transitions) [2], which uses an XUV attosecond pulse train (APT) to create an electron wave packet that will be later probed with the more intense IR field. The recorded photoelectron spectrum as a function of the time-delay (Δt) between the pulses captures the quantum interferences in the sideband reached through the two-photon paths: the (2n-1)th odd harmonic of the APT plus one IR-photon absorption and the (2n+1)th odd harmonic of the APT followed by an IR estimulated emission. In a structurless ionization region of an atom, the intensity of the sideband oscillates as a function of the time delay Δt such that $SB(\Delta t) = \alpha + \beta cos(2\omega_{IR}\Delta t - \Delta\phi - \Delta\theta)$, where $\Delta\phi$ is the relative phase between consecutive harmonics and $\Delta\theta$ is the atomic phase, i.e. the relative phase of the transition amplitudes for each path. Using well-characterized pulses, one can thus extract the atomic phase or, equivalently, the photoionization time delay $\tau = \Delta\theta/\omega$ [1]. While applied in the autoionization region of an atom, model preditions have also recently demonstrated the reconstruction of the coherent metastable wave packets [3]. In the present work, we employ a RABBIT scheme in H_2 to unravel the role of nuclear motion in the photoionization times in the presence of autoionization. Because electrons and nuclei share the total available energy, the photoelectron spectrum is no longer the relevant observable and one needs to analyze total energy paths, i.e., analysis of charged particles coincidence maps is required.

Experimentally, we employ a reaction microscope combined with an attosecond beamline [4]. We record kinematically complete momentum distributions of charged particles in coincidence, thus enabling to disentangle electrons associated to direct ionization from those originating from dissociating H_2 molecules. In order to be strongly resonant with the metastable doubly excited states (DES) of H_2, we also select counts originating from molecules with molecular axis aligned parallel to the XUV (and IR) polarisation direction. The molecular axis of H_2 molecules is retrieved from the direction of emission of the H^+ fragment within the axial recoil approximation. The preliminary experimental results are compared with theoretical predictions obtained illuminating the H_2 molecule with an APT formed by four XUV pulses of 400 as and energies around the 17th harmonic of a 780 nm generating IR field with 3×10^{11} W/cm^2 of intensity. These parameters are chosen to match the experimental conditions. The relative phases of the sidebands (see Fig. 1) strongly depend on the detuning of the photon energy from the metastable DES.

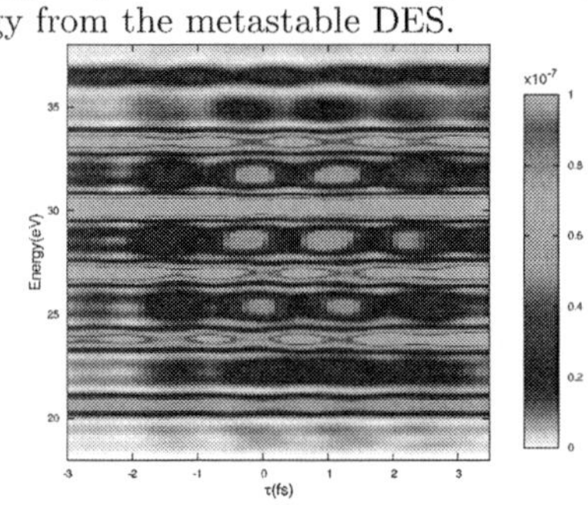

Figure 1. H_2 ionization signal (theory) for total absorbed energy as a function of time-delay between IR and XUV APT.

References

[1] J. M. Dahlström *et al* 2012 *JPB* **45** 183001

[2] P.M. Paul *et al* 2001 *Science* **292** 1689

[3] A. Jiménez-Galán *et al* 2014 *PRL* **113** 263001

[4] M. Sabbar, S. Heuser *et al* 2014 *Rev. Sci. Instrum.* **85** 103113

[1]E-mail: roger.bello@uam.es

Mapping ultrafast dynamics of highly excited D_2^+ by ultrashort XUV pump - IR probe laser schemes

R. Bello*[1], L. S. Martin†, C. W. Hogle†, A. Palacios*, J. L. Sanz-Vicario§, X. M. Tong‡, F. Martín*◇, M. Murnane†, H. C. Kapteyn† and P. Ranitovic†[2]

* Departamento de Química, M-13, Universidad Autónoma de Madrid, Cantoblanco 28049, Madrid, Spain
† JILA and Department of Physics, University of Colorado and NIST, Boulder, CO 80309-0440, USA
§ Grupo de Física Atómica y Molecular, Instituto de Física, Universidad de Antioquia, Medellín, Colombia
◇ Instituto Madrileño de Estudios Avanzados (IMDEA) en Nanociencia, Cantoblanco, 28049 Madrid, Spain
‡ Institute of Materials Science, University of Tsukuba, 1-1-1 Tennodai, Tsukuba, Ibaraki 305-8573, Japan

Synopsis An ultrashort XUV laser pulse ionizes the D_2 molecule into highly excited states of the ion that rapidly dissociates. These states are only populated through electron-electron interactions upon the one-photon absorption. A delayed interaction with an IR field leads to the Coulomb explosion of the molecule, whose nuclear fragments recorded in time map the dynamics of the ionic excited states ($2s\sigma_g$ and $2p\pi_u$). By varying the orientation of the light polarization, one can manipulate the molecular dynamics by modifying the relative contributions of the ionic states. Experimental and ab initio theoretical data are reported.

The ultimate goal of attosecond science is to achieve control of electron dynamics in matter. Our aim is to manipulate nuclear and electronic wave packets in excited [1] or ionized molecules [2]. The D_2 molecule is an ideal benchmark for coherent control mechanisms combining accurate experimental and theoretical methodologies. In our joint experimental and theoretical work, a single XUV pulse (energy 42.7 eV, 7 fs pulse duration) ionizes the D_2 molecule, creating a wave packet involving the highly excited states of D_2^+ ($2s\sigma_g$ and $2p\pi_u$). After a given delay, a 780 nm IR field ejects the second electron, leading to Coulomb explosion. All fragments are measured in coincidence using the well-known COLTRIMS technique, which allows for a complete reconstruction of the fragment momenta, and thus for identification of the molecular orientation. The nuclear kinetic energy distributions recorded in time map the evolution of the dynamics associated to highly excited ionic states, which are only being populated through electron-electron interactions after one-photon absorption. The background signal from the ground ($1s\sigma_g$) and first excited state ($2p\sigma_u$), on the other hand, is suppressed in the Coulomb explosion signal due to the large number of IR photons required to ionize D_2^+.

The coherent superposition of NWPs in the two electronic states $2s\sigma_g$ and $2p\pi_u$ can be manipulated by varying the UV pulse polarization with respect to the molecular axis. Moreover, the NWP associated to each ionic state accumulates a different momentum in each potential energy curve, therefore leading to nuclear fragments with different and identifiable energies in the Coulomb explosion channel. The agreement between theory and experiment is only achieved after performing advance calculations where nuclear and electron-electron correlation terms are accounted for, where one-photon ionizes and excited the molecule through a shake-up process. The dynamics of the highly excited ionic states, launched by the XUV pulse, is obtained by solving time-dependent Schrödinger equation using an $\mathcal{L}^2$ close-coupling method [3, 4], while the Coulomb explosion induced by the IR probe pulse is modeled as a simple sudden transition. Additionally, in the Coulomb explosion signal, we observe the quantum interference between the two sequential paths, associated to each intermediate ionic state.

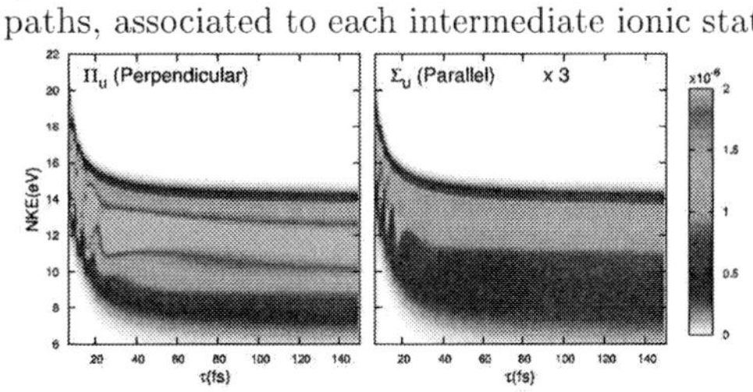

Figure 1. Calculated nuclear kinetic energy (NKE) distributions associated to the Coulomb explosion of D_2 as a function of the pump-probe time delay τ, for perpendicular ($^1\Pi_u$) and parallel ($^1\Sigma_u^+$) molecular orientations with respect to the polarization axis.

References

[1] P. Ranitovic *et al.* 2014 *PNAS* **111** 912

[2] F. Calegari *et al.* 2014 *Science* **346** 336

[3] A. Palacios *et al.* 2007 *PRA* **75** 013408

[4] J. L. Sanz-Vicario *et al.* 2006 *PRA* **73** 033410

[1]E-mail: roger.bello@uam.es
[2]E-mail: pranitovic@lbl.gov

P8

Finite element DVR method for molecular single and double ionization by strong laser pulses

Denis Jelovina[*1], **Johannes Feist**[◇], **Fernando Martín**[*†], **Alicia Palacios**[*2]

[*] Departamento de Química, Módulo 13,Universidad Autónoma de Madrid, Cantoblanco 28049, Madrid, Spain
[◇] Departamento de Física Teórica de la Materia Condensada and Condensed Matter Physics Center (IFIMAC), Universidad Autónoma de Madrid, Cantoblanco 28049, Madrid, Spain
[†] Instituto Madrileño de Estudios Avanzados en Nanociencia, Cantoblanco, 28049 Madrid, Spain

Synopsis We are developing an *ab initio* numerical method to compute molecular single and double ionization processes induced by intense laser pulses. We present our first time-dependent simulations in the H_2^+ molecular ion including both electronic and nucleardegrees of freedom. We compare different methods for the extraction of excitation and ionization probabilities from the numerical time-dependent wave function, which will be applied to more complex targets in the future.

We solve the time-dependent Schrödinger equation (TDSE) for the H_2^+ molecule in full dimensionality, with three spatial degrees of freedom for the electron, and nuclear motion restricted to one dimension (neglecting its angular dependence). The method is closely related to previous approaches employed in atoms [1, 2], employing a single-center expansion for the electronic degrees of freedom, with spherical harmonics describing the angular dependence. The radial component of both the electronic and nuclear motion is described using a finite-element discrete variable representation (FEDVR) basis.

In order to avoid unphysical reflections and to impose pure outgoing boundary conditions in the driven equations (see below), we apply exterior complex scaling (ECS) for both the electronic and internuclear radial coordinates [3].

As an initial benchmark result, we compute the (pure) dissociation and ionization cross sections for single-photon absorption, using three distinct methods. First, we obtain the total absorption (dissociation+ionization) from the optical theorem, $\sigma = \mathrm{Im}\langle\psi_0|\hat{\mu}\frac{1}{\hat{H}-\omega}\hat{\mu}|\psi_0\rangle$, where $|\psi_0\rangle$ is the ground state, $\hat{\mu}$ is the dipole operator (along the internuclear axis), $\hat{H}$ is the Hamiltonian, and ω is the driving frequency. Second, we propagate the system under the influence of an ultrashort laser pulse (which also gives access to multiphoton processes). We follow the formalism developed in [1] to implicitly integrate to infinite times for extracting the final observables. The dissocation and dissociative ionization cross sections are obtained by either integrating the probability flux through properly chosen surfaces, or by extracting the energy- and angle-resolved cross sections using surface integrals over testing functions [3, 1].

Figure 1 shows the cross sections for single-photon absorption using the latter implementation, extracted from the time-dependent wave packets obtained after interaction with laser fields of varying central frequency and duration. We obtain perfect agreement with ref. [4].

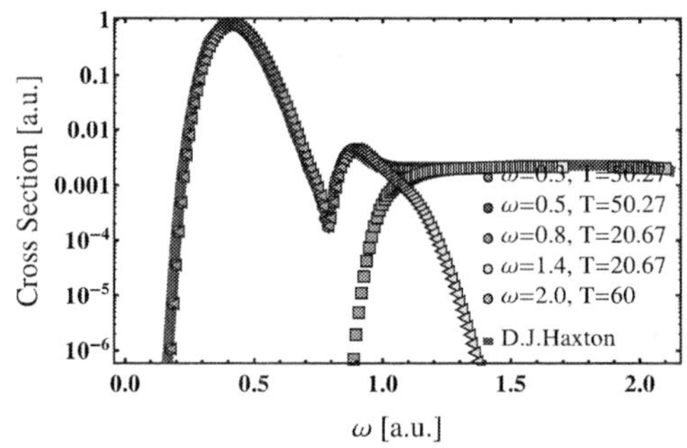

Figure 1. Absorption cross section for molecular ion as a function of the photon energy. Full line: Optical theorem. Rectangles: dissociative ionization. Triangles: dissociative excitation. Circles: total cross section. Each color stands for a different field use in the solution of the TDSE; the central frequency and pulse duration in atomic units are indicated.

[1]E-mail: denis.jelovina@uam.es
[2]E-mail: alicia.palacios@uam.es

References

[1] A. Palacios, T. N. Rescigno and S. W. McCurdy 2007 *Phys. Rev. A* **76** 043420

[2] J. Feist *et al.* 2008 *Phys. Rev. A* **77** 043420

[3] T. N. Rescigno and C. W. McCurdy 2000 *Phys. Rev. A* **62** 032706

[4] D. J. Haxton 2013 *Phys. Rev. A* **88** 013415

Merging quantum chemistry packages with B-splines for the multichannel scattering problem.

Carlos Marante*[1], Jesús González*, Inés Corral*, Markus Klinker*, Luca Argenti*, Fernando Martín*,†

* Departamento de Química, Módulo 13,Universidad Autónoma de Madrid, Cantoblanco 28049, Madrid, Spain
† Instituto Madrileño de Estudios Avanzados en Nanociencia (IMDEA-Nanociencia), Cantoblanco 28049, Madrid, Spain

Synopsis To study the photoionization of polyelectronic systems by attosecond pulses, we take advantage of existing quantum chemistry packages for the description of correlated electronic states, and of hybrid Gaussian-B-splines basis for the representation of continuum orbitals. In our approach, a short-range region, which can host all the interacting electrons and is described by commercial packages, is matched to a long-range region which describes single-ionization states in terms of a close-coupling expansion. We validate this approach showing multichannel ionization results for He.

Recent breakthroughs in ultrafast-laser technology, and more specifically the generation of extreme ultraviolet (XUV) and x-ray attosecond pulses [1, 2], have made it possible to study electron dynamics at its natural time scale. Since such pulses can ionize the atom or molecule with which they interact, the theoretical description of the observables in attosecond pump-probe experiments requires a good representation of the system's ionization continua in a wide energy range. While this is possible for very small systems such as He and H_2 [3–6], in more complex systems this is still a challenge, due to the complicate short-range structure of polyelectronic functions. Indeed, whereas sophisticated methods to compute the electronic structure of bound molecular states are implemented in commercial quantum chemistry packages (QCP) (e.g., MOLCAS [7]), comparable tools for the continuum are not really available.

To tackle this problem, we are developing a merge between existing QCPs and current numerical scattering methods. To do so, we separate the electronic configuration space in a short-range region, which can host all the interacting electrons, and a long-range region in which one outer electron interacts with a finite number of correlated parent ions. This outer electron is expressed in terms of a hybrid Gaussian-B-spline basis (GABS) which combines short-range Gaussian functions, compatible with standard QCPs, with B-splines [8], which are appropriate to represent the continuum.

To illustrate the viability of our approach, here we present the results for the multichannel ionization of He, which is a necessary step towards the study of more complex systems. The eigenphases above the N=2 threshold and the photoionization cross section we obtain are in excellent agreement with state-of-the-art benchmarks [9]. The good results obtained together with the great flexibility of QCPs position our method as a strong candidate for the theoretical study of the ionization of poly-electronic systems.

References

[1] M. Chini *et al* 2014 *Nature Photonics* **8**, 178-186

[2] T. Popmintchev *et al* 2010 *Nature Photonics* **4**, 822-832

[3] A. Jiménez-Galán *et al* 2014 *Phys. Rev. Lett.* **113** 263001

[4] C. Ott *et al* 2014 *Nature* **516**, 374-378

[5] G. Sansone *et al* 2010 *Nature* **465**, 763-766

[6] D. Dowek *et al* 2010 *Phys. Rev. Lett* **104** 233003

[7] G. Karlstrm *et al* 2003 *Comput. Mater. Sci.* **28** 2, 222-239

[8] C. Marante *et al* 2014 *Phys. Rev. A* **90** 012506

[9] L. Argenti and R. Moccia 2006 *J. Phys. B* **39** 2773

[1]E-mail: carlos.marante@uam.es

P10

Ultrafast electron dynamics in phenylalanine initiated by attosecond pulses

F. Calegari[*], D. Ayuso[†], A. Trabattoni[‡], L. Belshaw[¶], S. De Camillis[¶], S. Anumula[‡], F. Frassetto[§], L. Poletto[§], A. Palacios[†], P. Decleva[⊥], J. Greenwood[¶], F. Martín[†∇] and M. Nisoli[*‡]

[*] Institute of Photonics and Nanotechnologies, IFN-CNR, Piazza Leonardo da Vinci 32, 20133 Milano, Italy
[†] Departamento de Química, Módulo 13, Universidad Autónoma de Madrid, Cantoblanco 28049 Madrid, Spain
[‡] Department of Physics, Politecnico di Milano, Piazza Leonardo da Vinci 32, 20133 Milano, Italy
[¶] Centre for Plasma Physics, School of Maths and Physics, Queen's University Belfast, BT7 1NN, UK
[§] Institute of Photonics and Nanotechnologies, IFN-CNR, Via Trasea 7, 35131 Padova, Italy
[⊥] Dipartimento di Scienze Chimiche e Farmaceutiche, Università di Trieste, 34127, and CNR-IOM, Trieste, Italy
[∇] Instituto Madrileño de Estudios Avanzados en Nanociencia, Cantoblanco, 28049 Madrid, Spain

Synopsis We report the application of isolated attosecond pulses to prompt ionization of the amino acid phenylalanine and the subsequent detection of ultrafast dynamics on a sub–4.5-femtosecond temporal scale, which is shorter than the vibrational response of the molecule and can only be associated to purely electronic processes. Our interpretation is confirmed by the good agreement with numerical simulations in which the electronic wave packet generated by the attosecond pulse is quantitatively determined [1].

The process of electron transfer in complex molecules is of crucial importance since it triggers the first steps in a number of biological processes such as photosynthesis, cellular respiration or DNA damage [2]. Theoretical studies have shown that very efficient charge dynamics can be driven by purely electronic effects, which precede any rearrangement of the nuclear skeleton and can evolve on a temporal scale ranging from few femtoseconds down to tens of attoseconds [3,4]. This phenomenon has been referred to as charge migration [3].

Here we present an experimental demonstration of charge migration in a biologically relevant molecule, the amino acid phenylalanine [1]. Charge dynamics was initiated upon ionization by isolated sub-300-as pulses, with photon energy in the spectral range between 15 eV and 35 eV and probed by 4-fs, waveform-controlled near infrared (NIR) pulses, with central wavelength of 720 nm. The parent and fragment ions produced by the interaction with the pulses were collected by a linear time-of-flight device for mass analysis. The fragmentation yield of doubly charged immonium ion as a function of the time delay between the two pulses shows a clear oscillatory pattern which has been fitted to a sinusoidal function of period 4.3 fs. This ultrafast dynamics can only be associated with purely electronic effects, thus constituting the first experimental measurement of charge migration in a biomolecule.

We have performed theoretical calculations to describe the hole dynamics induced by the attosecond XUV pulse. Due to the large bandwidth of the pulse, a manifold of ionization channels is open, thus leading to a superposition of many ionic states, i.e., to an electronic wave packet. For all open channels, the ionization amplitudes were quantitatively determined using the static-exchange Density Functional Theory method [5]. The evolution of the electronic wave packet has then been described by employing a standard time-dependent density matrix formalism. The results of the numerical simulations clearly show the production of an ultrafast electron dynamics characterized by oscillation frequencies in good agreement with the experimental observations [1].

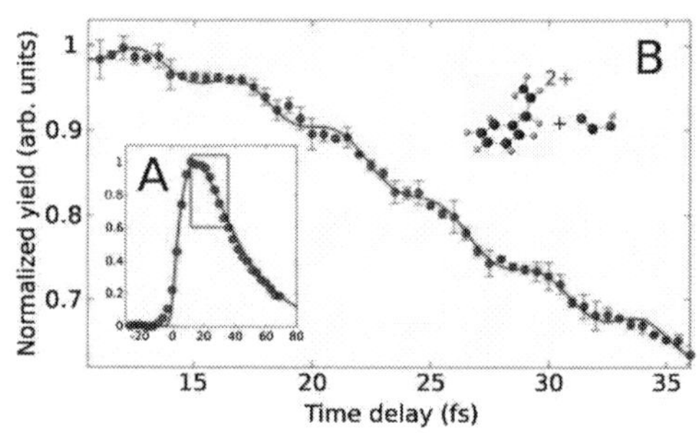

Figure 1. A) Fragmentation yield of doubly-charge immonium as function of the pump-probe delay. B) The same, measured with a 0.5-fs time step. The red line is the sum of the fitting curve shown in (A) and a sinusoidal function of period 4.3 fs.

References

[1] F. Calegari *et al,* 2014 *Science* **346** 6207 336
[2] J. Winkler *et al,* 2005 *Wiley-VCH* pp. 15-33
[3] L.S. Cederbaum *et al,* 1999 *Chem. Phys. Lett.* **307** 205
[4] F. Remacle *et al,* 2006 *P.N.A.S.* **103** 6793
[5] M. Stener *et al,* 2002 *Chem. Phys. Lett.* **351** 469

P11

Vibrationally resolved B 1s photoionization cross section of BF_3

D. Ayuso*[1], M. Kimura†, K. Kooser‡, M. Patanen¶, E. Plesiat*⊥, L. Argenti*, S. Mondal†, O. Travnikova¶§, K. Sakai†, A. Palacios*, E. Kukk‡, P. Decleva△, K. Ueda†, F. Martín*□, C. Miron¶

* Departamento de Química, Módulo 13, Universidad Autónoma de Madrid, Cantoblanco 28049, Madrid, Spain
† Institute of Multidisciplinary Research for Advanced Materials, Tohoku University, Sendai 980-8577, Japan
‡ Department of Physics and Astronomy, University of Turku, 20014 Turku, Finland
¶ Synchrotron SOLEIL, L'Orme des Merisiers, Saint-Aubin, BP 48, 91192 Gif-sur-Yvette Cedex, France
§ CNRS, UMR 7614, UPMC Univ Paris 06, LCPMR, 75005 Paris, France
⊥ Institut fur Physik, Humboldt-Universitat zu Berlin, Newtonstrasse 15, 12489 Berlin, Germany
△ Dipartimento di Scienze Chimiche e Farmaceutiche, Università di Trieste, 34127, and CNR-IOM, Trieste, Italy
□ Instituto Madrileño de Estudios Avanzados en Nanociencia, Cantoblanco, 28049 Madrid, Spain

Synopsis We present a study of the vibrationally resolved B 1s photoionization cross section of the BF_3 molecule. A combination of high-resolution photoelectron spectroscopy measurements and of state-of-the-art calculations shows the evolution of the photon energy dependence of the cross section from a complete trapping of the photoelectron wave (low energies) to oscillations due to intramolecular scattering [1, 2]. These diffraction patterns allow to access structural information of both the neutral molecule and the core-hole species generated upon photoabsoption [3].

The advent of third-generation synchrotron radiation facilities, in combination with high energy-resolution detection techniques, has opened the way for the investigation of vibrationally-resolved inner-shell photoionization in small molecules, where an electron is emitted from a 1s orbital of a first-row atom. We present a study on B 1s photoionization of BF_3, comparing experimental results with full first-principle calculations and showing that the most relevant features can be understood by means of simple models.

In order to calculate vibrationally-resolved cross sections, we have evaluated (bound and continuum) electronic wave functions using the static-exchange and the time-dependent DFT methods [4], developed by Decleva and collaborators, for different molecular geometries along the totally symmetric stretching mode. This mode is the most affected by the structural rearrangement accompanying core ionization [1, 2]. Our theoretical results are in good agreement with experimental measurements from SOLEIL synchrotron and, at high energies, qualitatively agree with a first-Born approximation model.

We have found that the relative cross sections show clear oscillations in the high-energy region as a function of photoelectron momentum which are due to an intramolecular scattering mechanism: in its way out of the molecule, the photoelectron is diffracted by the surrounding atomic centers, encoding the geometry of the molecule [1, 2, 3]. Very close to the photoionization threshold a complete trapping of the photoelectron is observed, manifesting itself as an emission angle dependent shape resonance feature [1, 2].

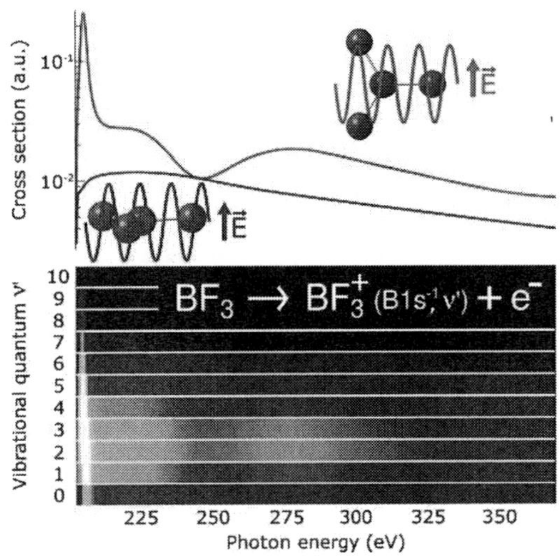

Figure 1. Calculated cross section (upper panel) for different molecular orientations and vibrationally resolved cross section in the totally symmetric stretching mode (lower panel) for the case of randomly oriented molecules.

References

[1] D. Ayuso *et al* 2015 *J. Phys. Chem. A* (Accepted)
[2] K. Ueda *et al* 2013 *J. Chem. Phys.* 139, 124306
[3] E. Kukk *et al* 2013 *Phys. Rev. A* 88, 033412
[4] H. Bachau *et al* 2001 *Rep. Prog. Phys.* 64, 1815

[1] E-mail: david.ayuso@uam.es

P12

Time delay anisotropy in photoelectron emission from the isotropic ground state of helium

S. Heuser[1], A. Jiménez-Galán[2 1], C. Cirelli[1], M. Sabbar[1], R. Boge[1], M. Lucchini[1], L. Gaallmann[1,4], I. Ivanov[5,6], A. Kheifets[5], J. M. Dahlström[7,8,9], E. Lindroth[7], L. Argenti[2], F. Martín[2,3,10], U. Keller[1]

[1] Physics Department, ETH Zurich, 8093 Zurich, Switzerland
[2] Departamento de Química, Módulo 13, Universidad Autónoma de Madrid, 28049 Madrid, Spain
[3] Instituto Madrileño de Estudios Avanzados en Nanociencia, Cantoblanco, 28049 Madrid, Spain
[4] Institute of Applied Physics, University of Bern, 3012 Bern, Switzerland
[5] Research School of Physics and Engineering,Australian National University, Canberra ACT 0200, Australia
[6] Center for Relativistic Laser Science, Institute for Basic Science (IBS), Gwangju 500-712, Rep. Korea, Asia
[7] Department of Physics, Stockholm University, AlbaNova University Center, SE-10691 Stockholm, Sweden
[8] Max Planck Institute for the Physics of Complex Systems, Noethnitzerstr. 38, 01187 Dresden, Germany
[9] Center for Free-Electron Laser Science, Luruper Chaussee 149, 22761 Hamburg
[10] Condensed Matter Physics Center (IFIMAC), Universidad Autónoma de Madrid, 28049 Madrid, Spain

Synopsis We show that time delays between electrons liberated from the $1s^2$ spherically symmetric ground state of helium depend on the emission direction of the electrons with respect to the polarization axis of the ionizing XUV light. This is attributed to the interplay between different final quantum states, which become accessible once two photons are involved in the photoionization process. This is a universal effect, which needs to be taken into account for any study dealing with photoionization dynamics.

The advent of attosecond science paved the way towards studying and understanding the nature of electron dynamics in atomic, molecular and condensed matter systems on their natural timescale. In particular, small delays in electron emission induced by single photon absorption have been measured with two different techniques: attosecond streaking and RABBITT[1]. Both methods are based on a pump-probe scheme, in which one pulse initiates the ionization process and the second pulse is used as a probe to retrieve the temporal information. So far, electron photoemission time delays have been measured between two different states of the same or different atoms. The dependence of the time delay on the electron emission direction, however, has not yet been investigated experimentally. When an electron is ionized by a single photon from an (n, ℓ) state, the dipole selection rules imply that the final photoelectron state has to be a $(E_f, \ell \pm 1)$ state. In the case of helium, starting from the $1s^2$ ground state, only one final state is available with a single photon absorption. However, if two photons are involved in the ionization process, as is the case in pump-probe schemes, two different final states are populated. The interplay between these final states may give rise to anisotropic time delays.

In this work, we report the first joint experimental and theoretical evidence of angle-dependent photoemission delays [2].Using the RABBITT technique, we measure a significant angular variation of the photoionization time delay of electrons ejected from the spherically symmetric $^1S^e$ ground state of helium, which can be as large as 50 attoseconds (Fig.1) for electrons emitted at an angle of 60 degrees with respect to the XUV polarization direction. The result is supported by virtually exact *ab initio* calculations which quantitatively reproduce the measured delays. We attribute the observed anisotropy to the interplay between the different quantum states that become accessible due to the two-photon transition process. Our observation gives insight into a new general aspect of attosecond measurements of electron dynamics. The universal effect studied in this work is potentially relevant for the interpretation of angular integrated experiments and has to be taken into account in any study dealing with electron dynamics including more complex systems such as e.g. molecules and condensed matter.

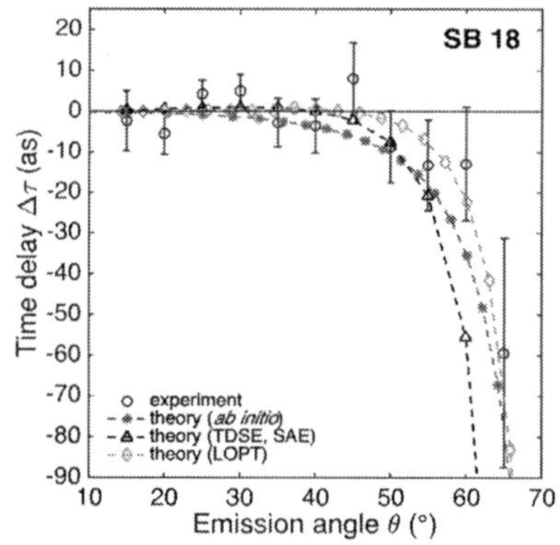

Figure 1. Angle-dependent photoemission time delay for photoelectrons with a kinetic energy of about 4.75eV.

References

[1] P.M. Paul *et al* 2001 *Science* **292** 1689

[2] S. Heuser *et al* 2015 arXive:1503.08966

[1]E-mail: alvaro.jimenez@uam.es

Phase Measurement of a Fano Resonance Using Tunable Attosecond Pulses

A. Jiménez-Galán[‡], M. Kotur[* 1], D. Guénot[*], D. Kroon[*], E. W. Larsen[*], M. Louisy[*], S. Bengtsson[*], M. Miranda[*], J. Mauritsson[*], C. L. Arnold[*], S. E. Canton[⋆], M. Gisselbrecht[*], T. Carette[†], J. M. Dahlstrom[†], E. Lindroth[†], A. Maquet[¶], L. Argenti[‡], F. Martín[‡◇§], A. L'Huillier[*]

[‡]Dep. Química, Módulo 13,Universidad Autónoma de Madrid, Cantoblanco 28049, Madrid, Spain
[*]Dep. Physics, Lund University, P. O. Box 118, SE-22100 Lund, Sweden
[⋆]Dep. Synchrotron Radiation Instrumentation, Lund University, P.O. Box 118, SE-22100 Lund, Sweden
[†]Dep. of Physics, Stockholm University, AlbaNova University Center, SE-10691 Stockholm, Sweden
[¶]Laboratoire de Chimie Physique-Matière et Rayonnement, UPMC, 75231 Paris, France
[◇]Condensed Matter Physics Center (IFIMAC), Universidad Autónoma de Madrid, 28049 Madrid, Spain
[§]IMDEA-Nanociencia, Cantoblanco, 28049 Madrid, Spain

Synopsis We study photoionization of argon atoms close to the $3s^23p^6 \rightarrow 3s^13p^64p$ Fano resonance using an attosecond pulse train and a weak infrared probe field. An interferometric technique combined with tunable attosecond pulses allows us to determine the phase of the photoionization amplitude as a function of photon energy. We interpret the experimental results using an analytical two-photon model based on the Fano formalism and obtain quantitative agreement.

Correlated electron dynamics induced by photoabsorption is a process of fundamental importance in nature. The development of table-top attosecond sources in the extreme ultraviolet (XUV) has opened up possibilities of accessing this dynamics directly in the time domain. In particular, correlation is responsible for the Auger decay, a resonant process that dramatically alters atomic photoionization spectra.

When a continuum state belonging to an open ionization channel interacts with a bound state belonging to a closed channel, autoionization may occur. Interference between the different ionization pathways leads to characteristic asymmetric Beutler-Fano profiles in the photoionization cross-section [1]. Studying this interaction in the time domain has been a major goal of attosecond science since the early days [2]. A method to do so is the so-called reconstruction of attosecond beating by interference of two-photon transitions (RABITT) technique [3], in which a train of XUV attosecond pump pulses is combined with a weak infrared (IR) probe with a variable time delay. The train, consisting of odd harmonics of the fundamental IR frequency, initiates the photoionization process. Further absorption or emission of an IR photon gives rise to sidebands. Due to the interference between two quantum paths, the sideband intensities oscillate as a function of the pump-probe delay. From the phase of these oscillations it is possible to obtain the phases of two-photon transition-amplitude and, in turn, of photoemission timing itself.

The RABITT technique has already been used to measure the effect of intermediate bound states on atomic two-photon transitions [4]. In this work, we present a study of the photoionization of argon using an interferometric method based on the RABITT technique, in the proximity of the $3s^23p^6 \rightarrow 3s^13p^64p$ autoionizing resonance, in which a tunable attosecond pulse train is used to initiate the photoionization process. As expected, changing the detuning between the resonance and the harmonic that is closest to it strongly affects the phases of the two adjacent sidebands. However, the phase profile clearly differs from the π jump observed for purely bound states. Our measurements are in quantitative agreement with the predictions of the analytical two-photon model in [5], which we extended to the multichannel case and parametrized with *ab initio* one-photon transition matrix elements obtained with a multi-configuration Hartree-Fock approach [6].

[1]E-mail: marija.kotur@fysik.lth.se

References

[1] U. Fano 1961 *Phys. Rev.* **124** 1866.
[2] M. Drescher *et al.* 2002 *Nature* **419** 803.
[3] P.M. Paul *et al.* 2001 *Science* **292** 1689.
[4] M. Swoboda *et al.* 2001 *Phys. Rev. Lett.* **104** 103003.
[5] A. Jiménez-Galán *et al.* 2014 *Phys. Rev. Lett.* **113** 263001.
[6] T. Carette *et al.* 2013 *Phys. Rev. A* **87** 023420.

P14

Temporal and spatial interferences in the laser-assisted photoionization of diatomic molecules.

Diego I.R. Boll*[1], Omar A. Fojón*[2]

*Laboratorio de Colisiones Atómicas, IFIR, CONICET-UNR, Rosario, Argentina

Synopsis We study theoretically the temporal and spatial interferences that may arise in the laser-assisted photoionization of diatomic molecules, particularly those leading to asymmetries in the final photoelectron distributions.

The development of extreme ultraviolet (XUV) light pulses with durations in the range of hundreds of attoseconds (as, 1 as $= 1 \times 10^{-18}$ s) has marked the begining of the attosecond physics [1]. The combination of these ligth pulses with a weak or moderately strong near-infrared laser field (NIR), as in the RABBITT or Streaking configurations, allow to monitor and/or control the electron dynamics.

In this work, several kinds of interference effects in laser-assisted photoionization of diatomic molecules are theoretically investigated by means of a Coulomb-Volkov model [5, 4]. In this model, the initial molecular wavefunctions are described as a linear combination of Slater type orbitals (STOs) variationally optimized whereas the final states are represented by separable Coulomb-Volkov type (SCV) wavefunctions [5].

The photoelectron angular distributions in the single-photon ionization reaction are compared to more elaborated results [6, 7]. In addition to the spatial interference effects arising from the coherent emission from both molecular centers, we analyze temporal interferences of the wavepackets released by a train of XUV pulses in the presence of laser bath [4]. Finally, the effect of the XUV pulse duration is analyzed [4].

In Fig. 1, we show the angle-energy photoelectron spectra for H_2^+ molecules ionized with XUV pulses of different duration in the presence of a NIR. As can be seen, asymmetric electron emission is obtained for XUV pulses with smaller durations (Fig. 1, first row), corresponding to the streaking regime. On the contrary, almost symmetric electron emission distributions are obtained when the XUV pulse duration is comparable to the period of the NIR (Fig. 1, second row).

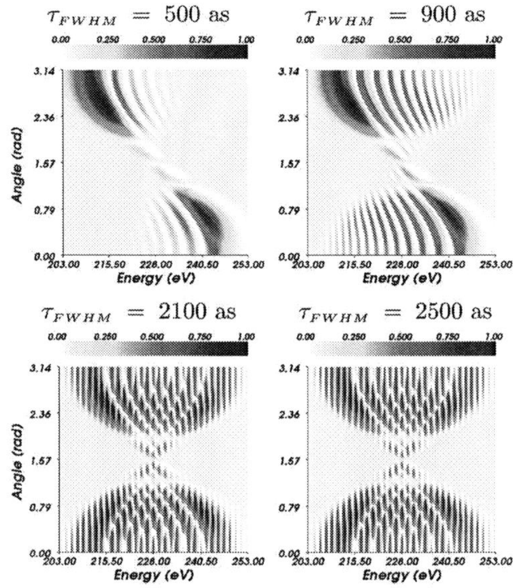

Figure 1. Angle-energy photoelectron spectra for H_2^+ ionized with XUV pulses of different durations, in the presence of a moderately intense NIR.

References

[1] F. Krausz, M. Ivanov 2009 *Rev. Mod. Phys.* **81** 163

[2] F. Kelkensberg *et al* 2011 *Phys. Rev. Lett.* **107** 043002

[3] G. Laurent *et al Phys. Rev. Lett.* **109** 083001

[4] D. Boll, O. Fojón 2014 *Phys. Rev. A* **90** 053414

[5] G. Yudin *et al* 2007 *J. Phys. B: At., Mol. Opt. Phys.* **40** F93

[6] X. Guan *et al* 2013 *Phys. Rev. A* **87** 053410

[7] J. Fernández *et al Phys. Rev. Lett.* **98** 043005

[1]E-mail: boll@ifir-conicet.gov.ar
[2]E-mail: fojon@ifir-conicet.gov.ar

P15

Molecular-frame photoelectron angular distributions for carbon 1s photoemission of methyl iodide

H. Fukuzawa[1,2] *, S. Yamada[1,2], Y. Ito[1,2], T. Tachibana[1,2], T. Takanashi[1,2], Y. Sakakibara[1,2], K. Nagaya[2,3], T. Nishiyama[2,3], T. Sakai[2,3], M. Yao[2,3], M. Oura[2], N. Saito[2,4], M. Stener[5], P. Decleva[6], and K. Ueda[1,2]

[1] Institute of Multidisciplinary Research for Advanced Materials, Tohoku University, Sendai 980-8577, Japan
[2] RIKEN SPring-8 Center, Sayo, Hyogo 679-5148, Japan
[3] Department of Physics, Kyoto University, Kyoto 606-8502, Japan
[4] National Metrology Institute of Japan, AIST, Tsukuba 305-8568, Japan
[5] Dipartimento di Scienze Chimiche e Farmaceutiche, Università di Trieste, Via L. Giorgieri 1, I-34127 Trieste, Italy
[6] Consorzio Interuniversitario Nazionale per la Scienza e Tecnologia dei Materiali Unita'di Trieste,34127 Trieste, Italy
[7] CNR-IOM DEMOCRITOS, 34149 Trieste, Italy

Synopsis We have investigated molecular-frame photoelectron angular distributions (MFPADs) for C 1s photoemission of methyl iodide (CH_3I) molecule experimentally and theoretically. In the case of polarization averaged MFPAD, we found that the photoelectron select directions of elements in the molecule depending on the photoelectron kinetic energies.

Photoelectron spectroscopy is one of the most widely used technique to investigate photoionization dynamics. In the case of isolated free molecules, it is important to measure photoelectron spectra in the molecular frame due to arbitrary molecular orientation. Recently, Williams *et al* measured molecular-frame photoelectron angular distribution (MFPAD) for C 1s photoemission from methane (CH_4) molecule averaged with respect to the orientation of light polarization and found that low energy electrons are emitted along C−H bond directions [1]. In this study we used methyl iodide (CH_3I) molecule as a target sample to investigate iodine substitute effect.

The experiment was carried out on the b-branch of the beamline BL17SU at SPring-8 [2]. The CH_3I vapor seeded by helium was introduced as a supersonic jet to the reaction chamber and crossed the linearly polarized photon beam at right angle. Three dimensional momenta of electrons and ions produced by soft x-ray irradiation of the sample were measured by the momentum-resolved electron−ion coincidence spectroscopy. This coincidence spectroscopy is based on recording the times-of-flight (TOFs) of electrons and ions with position and time-sensitive multihit-capable detectors (Roentdek HEX120 for electrons and HEX80 for ions). The TOF spectrometers are mounted face to face on the horizontal axis which is perpendicular to both, the photon beam and the molecular beam. A static extraction field and a uniform magnetic field are applied to the spectrometers so that all the electrons and ions ejected into the whole 4π sr solid angle are collected by the detectors. Knowledge of the arrival time and hit position of the particles on the detectors, (x, y, t), allows us to extract information on the linear momentum (p_x, p_y, p_z) for each particle. C−I axis of the CH_3I molecule is determined by measuring the momenta of CH_3^+ and I^+ in coincidence and used as the molecular axis.

The MFPADs of CH_3I molecule were measured at 3, 6, and 12 eV above C 1s ionization threshold. We found that the photoelectrons are mainly emitted in the direction of the C−H bond at 3 eV photoelectron energy, whereas 12 eV electrons are mainly emitted in the direction of the C−I bond. This trend was also confirmed by the time-dependent density functional theory calculation. The results suggest that the MFPADs have potential to be a tool to determine not only a geometry but also location of elements in free molecules.

* E-mail: fukuzawa@tagen.tohoku.ac.jp

References

[1] J. B. Williams *et al* 2012 *Phys. Rev. Lett.* **108** 233002

[2] H. Ohashi *et al* 2017 *AIP Conf. Proc.* **879** 523

P16

Subfemtosecond dynamics in dissociating core-excited CH_3I molecules studied with resonant Auger spectroscopy.

T. Marchenko[*,†,1], **G. Goldsztejn**[*,†], **L. Journel**[*,†], **R. Guillemin**[*,†], **O. Travnikova**[*,†], **A. F. Lago**[&], **D. Céolin**[#], **J.P. Rueff** [*,†,#], **R. Püttner**[§], **M. N. Piancastelli**[*,†,‡] **and M. Simon**[*,†,#]

[*] CNRS, UMR 7614, Laboratoire de Chimie Physique-Matière et Rayonnement, F-75005 Paris, France

[†] Sorbonne Universités, UPMC Univ Paris 06, UMR 7614, LCPMR, F-75005, Paris, France

[&] Centro de Ciências Naturais e Humanas, Universidade Federal do ABC, 09210-170, Santo André-SP, Brazil.

[#] Synchrotron SOLEIL, l'Orme des Merisiers, Saint-Aubin, BP 48, F-91192 Gif-sur-Yvette Cedex, France

[‡] Department of Physics and Astronomy, Uppsala University, SE-75120 Uppsala, Sweden

[§] Institut für Experimentalphysik, Freie Universität Berlin, D-14195 Berlin, Germany

Synopsis We present measurements of resonant Auger spectra in the CH_3I molecule near I 2p edge. We observe non-linear dispersion of the resonant spectator Auger lines indicating dissociative dynamics in the molecule on the sub-femtosecond time scale, which is determined by the natural lifetime of the I 2p core hole. We compare our observations to our earlier measurements in CH_3I using resonant inelastic x-ray scattering spectroscopy.

Ultrafast dissociation of inner-shell excited molecules was first observed in HBr molecule [1] and it has been since extensively studied in various molecular systems both theoretically and experimentally. Evolution of an excited molecule occurs on the time scale determined by the lifetime of the excited state. High-resolution spectroscopic techniques are presently largely recognized as important tools to study the mechanisms of relaxation and dissociation of inner-shell excited molecules. A number of studies in the soft X-ray range demonstrated fast molecular dissociation dynamics using Auger or X-ray emission spectroscopy [2]. In this photon energy range, inner shells of light elements can be excited, for example C 1s ~284 eV, O 1s ~543 eV. Extension of synchrotron radiation to higher photon energies in the "tender" X-ray range allows accessing deeper core shells in molecules and observing nuclear dynamics on increasingly shorter time scale providing information on the early stage of the chemical bond elongation. Using resonant inelastic X-ray scattering (RIXS) spectroscopy nuclear motion in the order of a few picometers within the time interval of ~1 fs was revealed in HCl molecule upon excitation of Cl 1s shell [3]. Very recently, we have further pushed the time limit and demonstrated a possibility to reveal the dynamical effects on 200 attosecond time scale in a core-excited CH_3I molecule excited at the I 2p shell [4].

Complementary to RIXS technique, high-resolution resonant Auger spectroscopy can also reveal subfemtosecond dynamical effects in molecules core-excited at the deep short-lived shells. High-energy Auger electron spectroscopy of core-excited gas-phase molecules has only recently become accessible at SOLEIL synchrotron. Excellent performance of the gas-phase measurements at the HAXPES end-station [5] at the GALAXIES beam line, providing photons in the "tender" X-ray range, has been demonstrated through a series of successful experiments [6-9].

Here we present measurements of Auger decay in CH_3I near the resonant excitation of the iodine L shell (~4.8 keV) to the lowest unoccupied molecular orbital. The Auger spectra are recorded for a series of photon energies, revealing various decay processes such as ultrafast dissociation, post-collision interaction effect, higher Rydberg transitions etc. Excellent instrumental resolution enabled measurements in the resonant Raman regime that provides the widths of the Auger spectral lines below the natural lifetime broadening of the core-excited states. The molecular dynamics occurring on the sub-femtosecond scale, determined by the core-hole lifetime, is identified through the characteristic non-linear dispersion and the width narrowing of the Auger spectral lines as a function of the photon energy. Our present resonant Auger measurements are compared to the earlier observations in CH_3I using RIXS spectroscopy. Our study demonstrates the potential of resonant Auger spectroscopy for application to the studies of sub-femtosecond dynamics in molecules excited at the deep core shells.

References

[1] P. Morin *et al* 1986 *Phys. Rev. Lett.* **56** 1913
[2] K. Ueda 2003 *J. Phys. B* **36**, 1
[3] M. Simon *et al* 2006 *Phys. Rev. A* **73**, 020706
[4] T. Marchenko *et al* 2011 *J. Chem. Phys.* **134** 144308
[5] D. Céolin, *et al* 2013 *J. Electron Spectrosc. Relat. Phenom.* **190**, 188
[6] M. Simon *et al* 2014 *Nature Comm.* **5** 4069
[7] M.N. Piancastelli *et al* 2014 *J. Phys. B* **47** 124031
[8] D. Céolin *et al* 2015 *Phys. Rev. A* **91** 022502
[9] R. Püttner *et al* 2015 *Phys Rev.Lett.* **114** 193001

[1] E-mail: tatiana.marchenko@upmc.fr

P17

Plasmon excitation due to Auger photoemission spectroscopy from Fe-Si alloys

Juana L. Gervasoni* [1], **Monika Jenko**† [2] and **Matjaž Godec**† [2]

*Centro Atomico Bariloche and Instituto Balseiro, Comision Nacional de Energia Atomica, S. C. de Bariloche, Rio Negro, Argentina. Member of the Consejo Nacional de Investigaciones Cientificas y Tecnicas (CONICET), Argentina.
† Institute of Metals and Technology, Lepi pot 11, 1000 Ljubljana, Slovenia.

Synopsis We investigate in detail the effects due to the interaction between an electron and a stationary positive ion (or atomic hole) in the neighborhood of a surface of Fe-Si, having a strong plasmon peak in their electron energy loss spectra. The plasmons leave a mark in the Auger and XPS spectrum which poses a challenge to the spectra analysis. The method described here is useful for studying multiple plasmon excitations and for understanding the electron spectra excited from these alloys.

Fe-Si has been the subject of experimental and theoretical interest for over fifty years [1]. The most direct way to shed light on the electronic structure, including many-body effects of a material is the resolved photoemission spectroscopy angle technique (ARPES, or AES). Furthermore, the energy resolution is not sufficient, in some cases, to give clear answers to the issue of the electronic nature of these states [2]. Therefore, highly resolved AES is used to study in detail the performance of single crystals of Fe-Si with a very low concentration of the impurity. In this paper we describe in detail the contribution to the energy loss of the external particles (electron plus holes) in the sample and the plasmon excitation due to this process.

We treat the Auger electron emission in a very simplified manner in order to be able to develop analytical expressions accounting for surface and bulk plasmon excitations. A detailed description of the photoelectron process was done in ref. [3]. Regarding the Auger electron emission, it will be schematized as a sudden creation of a static positive charge $q_+ = +2e$ and a moving negative charge $q_- = -e$, which is located at the same site as the positive charge at the moment of creation. This is justified by the fact that the Auger event happens very fast after the creation of the primary core-hole (in the order of 10^{-15} seconds).

[1] E-mail: gervason@cab.cnea.gov.ar
[2] E-mail: monika.jenko@imt.si
[2] E-mail: matjaz.godec@imt.si

In order to analyse the energy loss of Auger photoelectron due to plasmon excitations in Fe-Si alloys (incorporating a maximum of 5% Si), we consider internal emission for AES. We consider the emission of an electron with velocity u = 4 a. u. from a depth z_0 = 20 a. u. inside the alloy Fe-5%Si. We obtain the contributions of the electron and the holes to the total number of bulk (Q^B) and surface plasmons (Q^S). In Figure 1 we show the total average number of surface plasmons due to the presence of the electron as well as the holes.

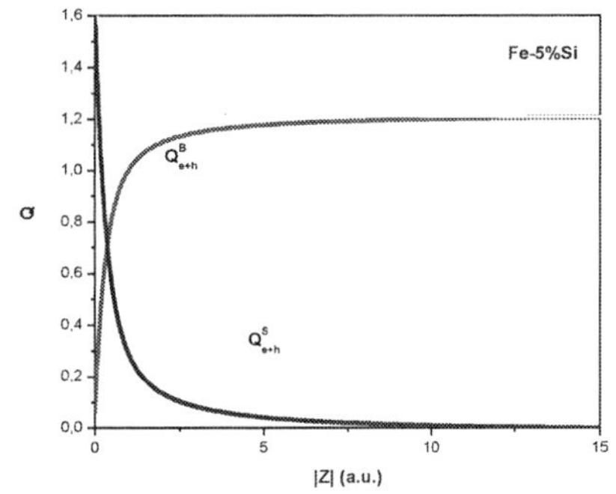

Figure 1. Average number of bulk (Q^B) and surface (Q^S) plasmon excitations (Q_B) in AES process, as a function of distance from the surface -z. We consider the emission of the electron with velocity u = 4 a.u., from a depth z_o = -20 a.u. inside the alloy. The hole is created in z_o, and remains static.

References

[1] Riseborough P S 2000, *Adv. Phys.* 49, 257–320.
[2] Jaccarino V, Wertheim G K, Wernick J H, Walker L R and Arajs S 1967, *Phys. Rev.* 160, 476–82.
[3] J. L. Gervasoni and N. R. Arista, *Surf. Sci.* 260, 329 (1992).

P18

Generalized Sturmian Function approach to the two-photon ionization of atoms

A.I. Gómez[†][1], G. Gasaneo[†], D.M. Mitnik[‡] and F.D. Colavecchia[§]

[†] Departamento de Física, Universidad Nacional del Sur - IFISUR-CONICET, 8000 Bahía Blanca, Buenos Aires, Argentina

[‡] Instituto de Astronomía y Física del Espacio, CONICET-UBA, C.C. 67, Suc. 28 (C1428ZAA), Ciudad Autónom de Buenos Aires, Buenos Aires, Argentina

[§] División de Coliciones Atómicas, Centro Atómico Bariloche, Av. Bustillo 9500, 8400 Bariloche, Argentina

Synopsis The main aim of this work is to apply the Generalized Sturmain Function (GSF) method to the study of the two-photon ionization of atoms. We present results obtained using a second order perturbation theory. The GSF used have the proper asymptotic behavior allowing us to extract the transition amplitudes from the obtained wave function.

The role of electron correlation in the few-photons processes has undergone experimental research due to the development of high-order harmonicsand free-electron laser (FEL)light sources. The process of one-photon doble ionization is today considered to be well understood as full agreement has been obtained between theory and experiments. However, the two-photon case is far more difficult to deal with, especially experimentally, due to the weak two-photon ionization yield. Non-sequential two-photon double ionization (TPDI) of helium is the simplest system where such processes occur. Its has been studied extensively during the last decade, both theoretically [1] and experimentally [2]. Nevertheless, there are still large discrepancies between the different theoretical approaches, and as such, the process is not considered to be fully understood. Before addressing the full He problem we will study the double ionization of Hydrogen. When working on a second order perturbation theory, the second order equation presents the particularity that the driven term is not vanishing at large distances. The aim of this contribution is to show how using the GSF method this problem can be easily tackled.

Our theoretical approach begins with the time-dependent Schrödinger equation. We apply the following successive steps: (1) a Fourier representation to characterize the time evolution of the system, (2) perturbative expansion, (3) a partial waves expansion and (4) the solution of the coupled radial equations using the GSF method [3]. These equations have the form:

$$\left[\omega_{ki} + \frac{1}{2}\frac{\partial^2}{\partial r^2} - \frac{l(l+1))}{2r^2} + \frac{Z}{r}\right]\varphi_l^{(n)}(r,\omega) = f_{RHS}^{(n)}(r,\omega)$$

where $\varphi_l^{(n)}(r,\omega)$ is the radial solution to each order and $f_{RHS}^{(n)}(r,\omega)$ is the driven term that depends on $\varphi_l^{(n-1)}(r,\omega)$ and on the field. The GSF used to solve the equations corresponding to the first and second orders are constructed possessing the appropriate asymptotic conditions. The treatment of the second order equation should be particular due to the fact that its driven term does not vanishes when $r \to \infty$ as can be seen on Fig 1.a. This non-vanishing behavior enforce on the second order wave function a "beat" type asymptotic behavior as shown in Fig. 1b. GSFs can be easily constructed possessing the "beat" behavior (see Fig. 1.c). When solving the second order equation using this basis set convergence is naturally increased and they allow the extraction of the transition amplitude directly from it asymptotic behavior of the wave function.

The application of the same methodology to the He atom will be discussed on the presentation.

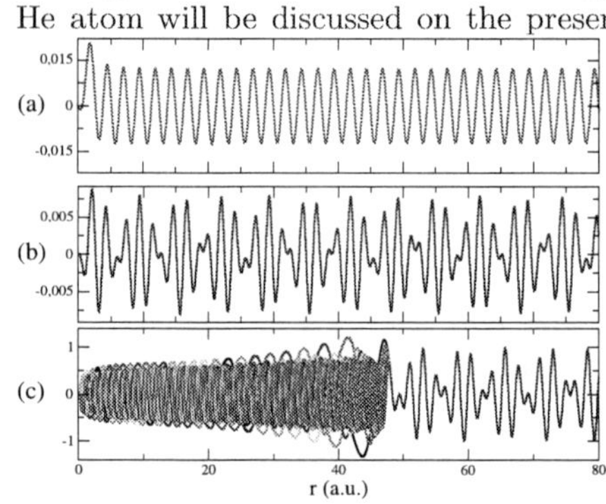

Figure 1. (a) Second order driven term. (b) Second order scattering wavefunction. (c) A suitable set of Generalized Sturmian Functions.

[1] E-mail: unilo.2099@gmail.com

References

[1] L.Malegat *et al.* 2012 *J. Phys. B* **45** 175601

[2] Sh.A. Abdel-Naby *et al.* 2013 *Phys. Rev. A* **87** 063425

[3] G. Gasaneo *et al.* 2013 *Adv. Quantum Chem.* **67** 153.

P19

Photoionization of CH_4, H_2O and NH_3: a Sturmian approach

C. M. Granados–Castro*,† 1, L. U. Ancarani*, G. Gasaneo†,‡, D. M. Mitnik§,‡

* Équipe TMS, SRSMC UMR 7565, Université de Lorraine, 57078 Metz, France

† Departamento de Física, Universidad Nacional del Sur, 8000 Bahía Blanca, Argentina

§ Instituto de Astronomía y Física del Espacio (IAFE) and Universidad de Buenos Aires, Argentina

‡ Consejo Nacional de Investigaciones Científicas y Técnicas (CONICET), Argentina

Synopsis A Sturmian approach, based on Generalized Sturmian Functions, is used to study single photoionization of small molecules. A molecular model potential, that plays the role of a scattering potential, is proposed to describe the interaction of the ejected electron with the parental ion. The random orientation of the molecule is taken into account through an angular average. Calculated photoionization cross sections for CH_4, NH_3 and H_2O are reported.

In most cases, the Hamiltonian that describes molecular systems is highly non-central and has a multicenter nature. Thus it is not easy to find an accurate representation of continuum states of molecules [1]. The Sturmian approach [2], using Generalized Sturmian Functions (GSF), has been applied successfully for the study $(e, 2e)$ and $(e, 3e)$ [3] and $(\gamma, 2e)$ [4] in He, but the extension of the method to molecular systems is under development [1]. In this contribution, we implement the Sturmian approach to single photoionization (PI) of small molecules.

In our study of PI processes we use: (i) the single active-electron approximation; (ii) a non-central molecular model potential [5] as scattering potential, (iii) the one-electron wave functions in the one-center expansion calculated by Moccia [6]; and (iv) an angular average (AA) over all possible spatial orientations of the molecule. Such AA is performed in order to take into account the random orientation of the molecule in the laboratory frame. This is done either with an AA of the molecular model potential (pre-averaged calculation), or using a non-central model potential and then evaluating the angular average of the calculated cross sections (post-averaged).

To calculate the PI cross sections, we solve the time-independent, first-order perturbative, Schrödinger equation, expanding the scattering wave function in a GSFs basis set. Their adequate asymptotic behavior allows us to extract directly the transition amplitudes from the expansion coefficients [1], and then no further integrals are required. The use of a non-central model potential gives a set of angular-coupled driven equations. The calculated cross sections (partially presented in [1]) are compared with theoretical and experimental data. In Fig. 1 we provide an example.

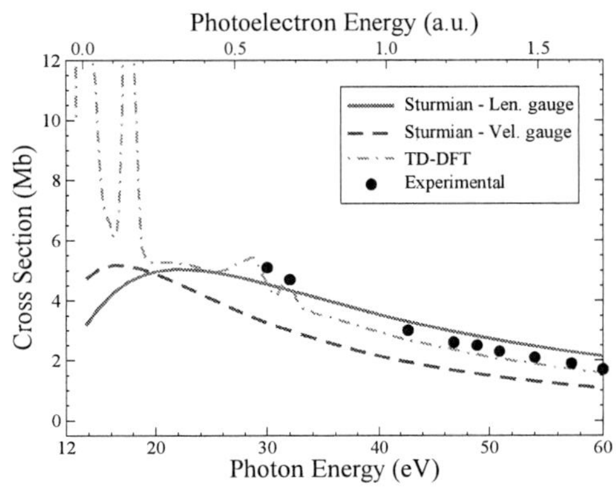

Figure 1. PI cross section in Mbarns for the single ionization of the $1b_1$ orbital of H_2O, using a non-central model potential. Our calculations in length (red, solid) and velocity (blue, dash) gauges are compared with the TD-DFT calculation [7] (green, dash-dot) and with experimental data [8].

[1]E-mail: carlos.mario-granados.castro@univ-lorraine.fr

References

[1] C. M. Granados–Castro *et al* 2015 *J. Phys: Conf. Series* **601** 012009; C. M. Granados–Castro *et al* 2015, submitted to *Adv. Quantum Chem.*

[2] G. Gasaneo *et al* 2013 *Adv. Quantum Chem.* **57** 153

[3] M. J. Ambrosio *et al* 2015 *J. Phys. B: At. Mol. Opt. Phys.* **48** 055204

[4] J. M. Randazzo *et al* 2015, submitted to *Eur. J. Phys. D*

[5] L. Fernández–Menchero and S. Otranto 2010 *Phys. Rev. A* **82** 022712

[6] R. Moccia 1964 *J. Chem. Phys.* **40** 2164; **40** 2176; **40** 2186

[7] M. Stener *et al* 2002 *Chem. Phys.* **282** 337

[8] M. S. Banna *et al* 1986 *J. Chem. Phys.* **84** 4739

P20

Electron impact ionization of CH_4, H_2O and NH_3: a Sturmian approach

C. M. Granados–Castro[*,† 1], **L. U. Ancarani**[*], **G. Gasaneo**[†,‡], **D. M. Mitnik**[§,‡]

[*] Équipe TMS, SRSMC UMR 7565, Université de Lorraine, 57078 Metz, France
[†] Departamento de Física, Universidad Nacional del Sur, 8000 Bahía Blanca, Argentina
[§] Instituto de Astronomía y Física del Espacio (IAFE) and Universidad de Buenos Aires, Argentina
[‡] Consejo Nacional de Investigaciones Científicas y Técnicas (CONICET), Argentina

Synopsis We implement a Sturmian approach, based on the Generalized Sturmian Functions method, for the study of single electron impact ionization of small molecules. Molecular model potentials are proposed to describe the interaction between the ejected electron with the parental ion. Calculated triple differential cross sections for CH_4, NH_3 and H_2O are reported. The random orientation of the molecule is taken into account through an angular average.

Since the Hamiltonian for molecules is generally multicenter and highly non-central, it makes life harder to study $(e, 2e)$ processes than in the case of atomic targets. The Sturmian approach, using Generalized Sturmian Functions (GSFs) [1], has been applied to study $(e, 2e)$, $(e, 3e)$ [2] and $(\gamma, 2e)$ [3] in He. Recently, it has been used to investigate single photoionization of atoms and molecules [4]. We present here the implementation and application to ionization of molecules by electron impact.

Within our Sturmian approach we use (i) the single-electron and the frozen-core approximations; (ii) the one-electron bound wave functions in the one-center expansion, calculated by Moccia [6]; (iii) a plane wave to describe both initial and final states of the projectile; (iv) a scattering potential given by a non-central molecular model potential [5] or its angular average (central potential); and (v) an expansion in one-particle GSF basis for the scattering wave function.

Since the molecular spatial orientation cannot be resolved in most experiments, an angular average (AA) over all possible spatial orientations of the molecule must be performed [4]. Such averaging can be done either with an AA of the molecular model potential, or using a non-central model potential and then evaluating the AA of the resulting transitions amplitude (TA).

To obtain the required TAs, we follow the same procedure as given in [4] for photoionization, expanding the scattering wave function in a GSFs basis and solving the resulting angular-coupled system of driven equations. The asymptotic behavior of GSFs allows us to extract directly the TAs from the expansion coefficients [4].

The calculated triple differential cross sections will be compared with theoretical and experimental data for CH_4, H_2O and NH_3. In Fig. 1 an example is provided.

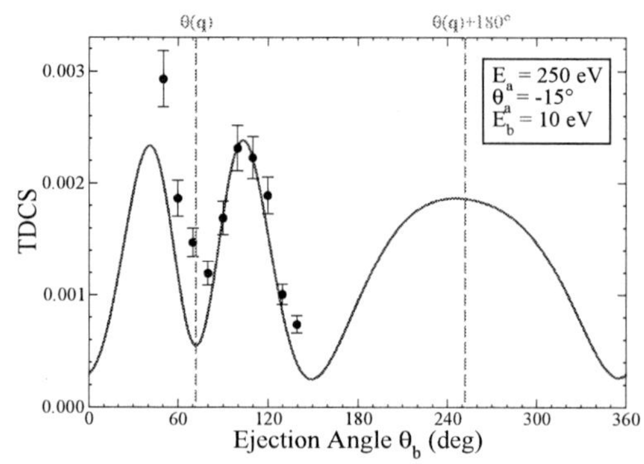

Figure 1. Triple differential cross section (relative scale, blue solid) for the single ionization of the $1b_1$ orbital of H_2O, obtained after an AA over spatial orientation. It is compared to relative experimental data [7].

References

[1] G. Gasaneo *et al* 2013 *Adv. Quantum Chem.* **57** 153

[2] M. J. Ambrosio *et al* 2015 *J. Phys. B: At. Mol. Opt. Phys.* **48** 055204

[3] J. M. Randazzo *et al* 2015, submitted to *Eur. J. Phys. D*

[4] C. M. Granados–Castro *et al* 2015 *J. Phys: Conf. Series* **601** 012009; C. M. Granados–Castro *et al* 2015, submitted to *Adv. Quantum Chem.*

[5] L. Fernández–Menchero and S. Otranto 2010 *Phys. Rev. A* **82** 022712

[6] R. Moccia 1964 *J. Chem. Phys.* **40** 2164; **40** 2176; **40** 2186

[7] D. Milne–Brownlie *et al* 2004 *Phys. Rev. A* **69** 032701

[1]E-mail: carlos.mario-granados.castro@univ-lorraine.fr

P21

Convergent Close Coupling double ionization amplitude extraction with Hyperspherical Sturmian Functions

M.J. Ambrosio*[1], D.M. Mitnik*, G. Gasaneo†, J.M. Randazzo°, I. Bray‡ and L.U. Ancarani+

* Instituto de Astronomía y Física del Espacio (IAFE, CONICET–UBA).
† Departamento de Física, Universidad Nacional del Sur and CONICET, 8000 Bahía Blanca, Buenos Aires, Argentina.
°División Física Atómica, Molecular y Óptica, Centro Atómico Bariloche and CONICET, 8400 S. C. de Bariloche, Río Negro, Argentina.
‡ARC Centre for Antimatter-Matter Studies, Curtin University, GPO Box U1987, Perth, WA 6845, Australia.
+Théorie, Modélisation, Simulation, SRSMC, UMR CNRS 7565, Université de Lorraine, 57078 Metz, France.

Synopsis The Convergent Close Coupling method is an *ab initio* scheme which usually defers the explicit wave function calculation in favor of the transition matrix. We show that, for a three-body ionization problem, the method is capable of yielding a wave function equivalent to that obtained by other *ab initio* approaches, such as the Generalized Sturmian Functions method. From the asymptotic regime of the wave function one may recover, in an alternative way, the transition matrix.

The Convergent Close Coupling (CCC) method was introduced in the early 1990s to solve atomic collision three-body problems [1]. It does not require an explicit evaluation of the resulting wave function, proceeding directly with the T-matrix.

In this contribution we show that the CCC method is indeed able to yield a scattering wave function as shown in Fig. 1 for the Hydrogen ionization problem. We have compared it with a counterpart calculated with the Generalized Sturmian Functions method [2]. As we found a good match, they carry globally the same physical information.

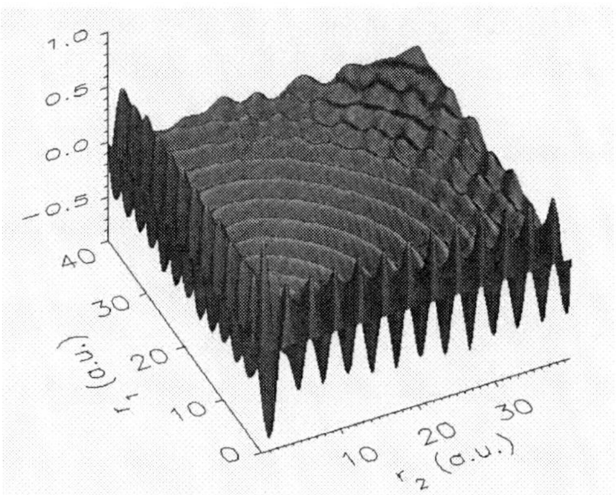

Figure 1. Real part of the CCC scattering solution for the Temkin-Poet ionization of Hydrogen by electron impact.

In order to extract the scattering information from the CCC wave function, we use here a hyperspherical three–body basis containing Hyperspherical Generalized Sturmian Functions (HGSF), reviewed in Ref. [2]. As few as 30 hyperangular functions and 18 hyperradial ones are needed to reexpand the CCC scattering function up to $\rho = \sqrt{r_1^2 + r_2^2} = 35$ a.u. With such a HGSF expansion one can easily extract the transition matrix from the true asymptotic regime (Fig. 2). Cross sections can then be computed in a straightforward way - from the CCC scattering function - with the same technique used recently for the double ionization of Helium [3, 4].

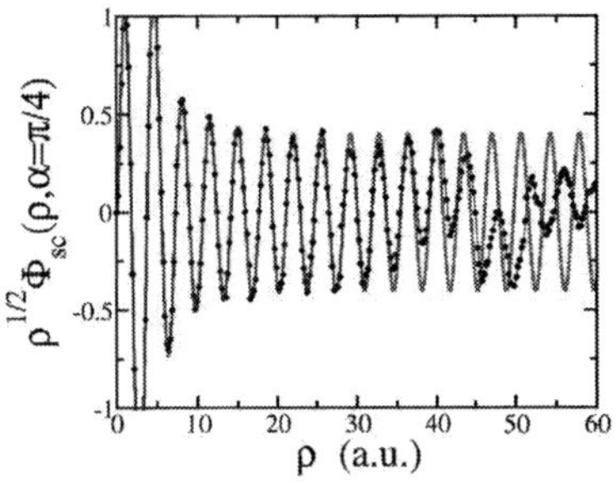

Figure 2. Real part of the CCC scattering function along $r_2 = r_1$ ($\alpha = \pi/4$) (thin black with black circles) and its reexpansion in the HGSF basis (red continuous line).

References

[1] I. Bray *et al* 2002 *J. Phys. B* **35** R117

[2] G. Gasaneo *et al* 2013 *Adv. Quant. Chem.* **67** 153-216

[3] M.J. Ambrosio *et al* 2015 *Phys. Rev. A* **91** 012704

[4] M.J. Ambrosio *et al* 2015 *J. Phys. B* **48** 055204

[1]E-mail: mja1984@gmail.com

P22

Investigation of triple differential cross section for electron impact ionization of methane molecule

Mevlut Dogan*[1], Murat Yavuz*, Zehra Nur Ozer*, Nimet Isik*, Semiha Bahceli‡, Adnan Naja†,

* Department of Physics, e-COL Laboratory, Afyon Kocatepe University, Afyon, Turkey
‡ Department of Physics, Süleyman Demirel University, Isparta, Turkey
† Lebanese University, LPM, EDST, Tripoli, Lebanon

Synopsis We present experimental results of triple differential cross sections (TDCS) for $1t_2$ outer orbital of methane molecule. Angular distributions of ejected electrons have been measured after ionizing electron collisions with methane at primary energy of 250 eV using a conventional electron spectrometer. Measurements are taken for 10° scattering angle at 30 eV ejected electron energy.

Many applications in radiological, atmospheric and plasma physics require differential ionization cross sections over a broad range of electron energies. To meet this need it is useful to have experimental data from diverse sources to unify theoretical models and to provide reliable estimates of cross sections. On the other hand, successes in research of cross sections for molecular structures are quite limited mostly because of experimental difficulties. In order to measure reliable cross sections it is necessary to use an electron spectrometer which is capable to measure under asymmetric kinematics.

In this work, methane molecule was chosen because that it is the smallest hydrocarbon and so is a relatively simple polyatomic molecule. The simple electronic structure has made methane an ideal prototype to model organic systems and so it has often been employed when developing models of biological matter or planetary atmospheres.

The apparatus developed for coincidence measurements of ejected and scattered electrons. Triple differential cross section measurements were performed for outer $1t_2$ orbital at 250 eV incident and 30 eV ejected electron energies (Figure 1).

The results reported in this study will present data for developing theoretical models to molecular dynamics.

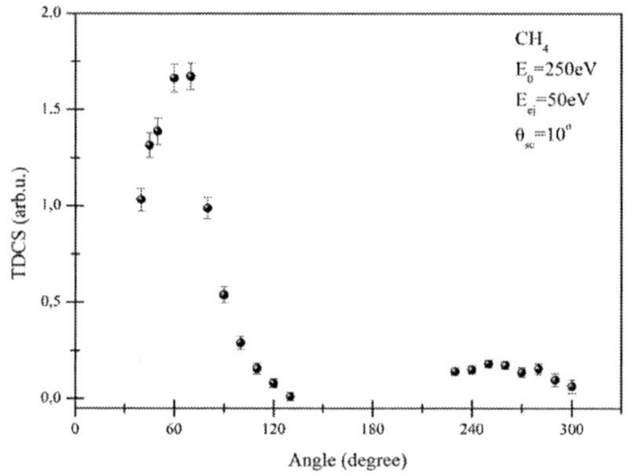

Figure 1. TDCS of methane molecule for incident electron energy 250 eV, 10° scattering angle and ejected electron energy 50 eV.

Acknowledgments
This work was supported by the Scientific and Technological Research Council of Turkey (TUBITAK) through grant 109T738 and AKU-BAPK through grant 14.HIZ.DES.80.

[1]E-mail: mdogan@aku.edu.tr

P23

(e, 2e) coincidence studies for simultaneous excitation–ionization to $He^+(n=2)$ by electron impact

Mevlut Dogan*[1], Albert Crowe[†], Oleg Zatsarinny[‡], Klaus Bartschat[‡2]

[*] Department of Physics, e-COL Laboratory, Afyon Kocatepe University, Afyon, Turkey
[†] School of Chemistry, Newcastle University, Newcastle, United Kingdom
[‡]Department of Physics and Astronomy, Drake University, Des Moines, Iowa 50311-4505, USA

Synopsis Experimental and theoretical triple-differential cross sections are presented for electron-impact ionization of helium, leaving the residual ion in the $He^+(n=2)$ states. We consider incident-electron energies between 200 eV and 635 eV, scattering angles of -4° and -11°, and an ejected-electron energy of 10 eV.

Electron collisions with atomic and molecular targets and their ions are an important area of atomic physics, both experimentally and theoretically. Such processes provide a means of investigating the dynamics of few-body systems at a fundamental level. Electron-helium scattering, in particular, is of interest in diverse applications and to basic collision theory as the prototype of a Coulomb four-body problem.

Simultaneous ionization–excitation of helium from its ($1s^2$) ground state by electron impact is interesting, because three electrons change their quantum state significantly during such a collision. The cross-sections are relatively small, usually two to three orders of magnitude smaller than for ionization without excitation. The low signal places serious challenges not only to experiment, but also to theoretical efforts, since the results tend to become very sensitive to small changes in any theoretical model.

We present triple-differential cross sections (TDCS) for simultaneous electron-impact ionization–excitation of $He(1s^2)$, with the residual ion left in the $He^+(n=2)$ state. Measurements and calculations were performed for incident-electron energies of 200 eV, 570 eV, and 635 eV, scattered-electron angles of -4° and -11°, and an ejected-electron energy of 10 eV. While the TDCS for ionization without excitation exhibits the characteristic behavior of peaks along and opposite to the momentum-transfer direction, the corresponding peaks are well removed from these directions if the ion is left in an excited state. In some cases, the recoil peak is larger than the binary peak. Furthermore, the angular dependence of the TDCS tends to become strongly dependent on the scattering angle for fixed incident and ejected electron energies and on the incident energy for fixed scattering angle and ejected electron energy – more so than for ionization without excitation.

Figure 1 shows an example of our results. The experimental data are compared with theoretical predictions from first-order and second-order hybrid distorted-wave plus R-matrix models (DWB1-RM and DWB2-RM, respectively). Although the experimental data are not absolute, it is clear that the angular dependence is better described by the DBW2-RM than by the DBW1-RM model. Predictions from a fully nonperturbative B-spline R-matrix approach will be presented at the conference.

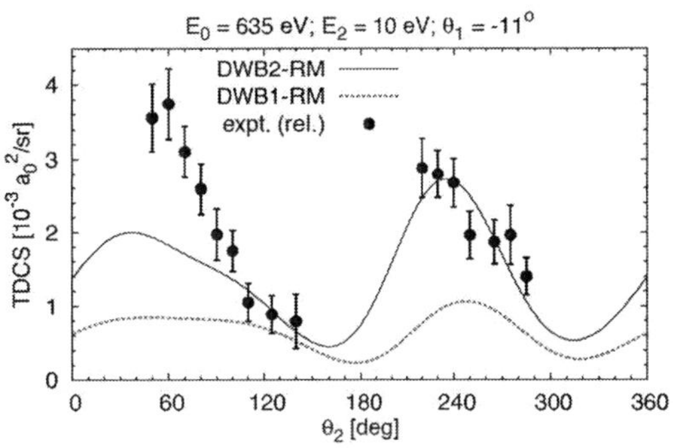

Fig. 1. TDCS for electron impact ionization of $He(1s^2)$, with the residual ion left in the $He^+(n=2)$ state. The experimental data were visually normalized to the DWB2-RM results.

This work was supported, in part, by AKU-BAPK through grant 14.HIZ.DES.80 and by the United States National Science Foundation under grants PHY-1212450 and PHY-1430245, and the XSEDE allocation PHY-090031.

[1]E-mail: mdogan@aku.edu.tr

[2]E-mail: klaus.bartschat@drake.edu

P24

Significance of projectile scattering angle on the observation of Young's type interference effects for diatomic molecules

Zehra Nur Ozer*[1], Hari Chaluvadi†, Mevlut Dogan*, Don Madison†

* Department of Physics, e-COL Laboratory, Afyon Kocatepe University, 03200, Afyon, Turkey
† Department of Physics, Missouri University of Science and Technology, Rolla, Missouri, USA

Synopsis We have found experimentally and confirmed theoretically that, in the angular scan for ejected electrons, Young's type interference effects can be seen in the triple differential cross sections (TDCS) for the diatomic molecules H_2 and N_2 at intermediate energies.

The quantum equivalent of Young's two slit experiment in optics would be charged particle scattering from diatomic molecules. Cohen and Fano [1] were the first to predict that coherent emission from two molecular centers would produce interference effects which depends on the wavelength of the ejected electrons compared to internuclear distance for a diatomic molecule [1]. Stimulated by this pioneering work, some experimental and theoretical studies investigated possible interference effects for electron impact ionization of diatomic molecules [2-6]. The best way to see interference effects is to examine the ratio between molecular cross sections to atomic cross sections of two isolated atoms which should contain no two center interference effects.

Considering the Cohen and Fano model for interference phenomena, the most important variable is expected to be the wavelength of the ejected electron. Consequently, we first searched for interference effects for electron impact ionization of the H_2 molecule by changing the ejected electron energy (wavelength) [7,8]. However, in a recent study, we showed that there are three types of possible two-center interference effects and the most important one is the diffraction of the projectile from two scattering centers for H_2 and N_2 molecules [9-11].

We have developed both a theoretical and experimental framework to characterize interference phenomena and have applied it to H_2 and N_2 molecules at intermediate energies. The experimental data presented in this work are taken using a conventional electron spectrometer that can operate over a range of scattering angles and energies for coplanar asymmetric geometries.

E-mail: zehraerengil@aku.edu.tr

Theoretical cross section calculations are performed using the Molecular Three Distorted Wave approximation (M3DW) where it is possible to isolate different physical effects in the calculation.

We will discuss the importance of scattered and ejected electron angles on interference effects for the diatomic molecules H_2 and N_2 at intermediate electron energies.

Acknowledgments:

This work was supported by AKU, BAPK through grants 14.FENED.06 and 14.HIZ.DES.80. The theoretical work of SA and DM was supported by US National Science Foundation under Grant. No. PHY-1068237.

References

[1] H. D. Cohen, U. Fano, 1966 *Phys. Rev.* **150(1)** 30.
[2] C. R. Stia *et al* 2003 *J. of Phys. B* **36(17)** L257.
[3] D. S. Milne-Brownlie *et al* 2006 *Phys. Rev. Lett.* **96(23)** 233201.
[4] E. S. Casagrande *et al* 2008 *J. of Phys. B* **41(2)** 025204.
[5] D. Misra *et al* 2006 *Physical Review A*, ***74*(6), 060701.**
[6] N. Stolterfoht *et al* 2001 *Physical Review Letters*, ***87*(2), 023201.**
[7] D. Milne-Brownlie *et al* 2006 *Physical Review Letters*, ***96*(23), 233201.**
[8] E. S. Casagrande *et al* 2008 *Journal of Physics B: Atomic, Molecular and Optical Physics*, ***41*(2), 025204.**
[9] Z. N. Ozer *et al* 2013 *Physical Review A*, ***87*(4), 042704.**
[10] H. Chaluvadi *et al* 2015 submitted to *J. Phys. B: At. Mol. Opt. Phys.*
[11] Z N.Ozer et al., 2015, preparing for publication.

P25

Investigation of electron impact double excitation of the autoionizing states of Helium by (e, 2e) experiments

Melike Ulu*[1], Omer Sise[§], Ali Alpergun*, Nurcin Karadeniz*, Mevlut Dogan* and Albert Crowe[†]

[*] Department of Physics, e-COL Laboratory, Afyon Kocatepe University, Afyon, Turkey
[§] Department of Science Education, Faculty of Education, Suleyman Demirel University, Isparta, Turkey
[†]School of Chemistry, Newcastle University, Newcastle, United Kingdom

Synopsis We present experimental measurements of triple differential cross sections (TDCS) for the helium autoionizing resonance states. Triple differential cross-sections can be obtained by measuring the energy and angular distributions of outgoing electrons by use traditional electron spectrometer.

Autoionization in atomic collisions is a fundamental process that has attracted a lot of interest for several decades since Madden and Codling conducted their pioneering work on helium in this field [1].

Autoionizing states are degenerate with continuum (final ion plus ejected electron) states from the direct ionization process. The dynamics of this process can be uniquely determined by (e, 2e) experiments in which both the outgoing electrons resulting from the same ionization event are detected in time coincidence.

The (e, 2e) technique investigates electron impact ionization provides the most detailed insight into the reaction of many-body systems. However, due to the low cross sections and difficulties in achieving the required energy resolution, the experimental investigations of the double excitations of helium by this technique are very limited. In two (e,2e) papers [2,3], our group has published on He *2l2l'* autoionizing states for incident electron energy of 250 eV at coplanar geometry. These works show that the resonance contributions together with the strong background of direct ionization processes lead to a complicated asymmetric structure of the resonance profile in the TDCS.

In the conference, we will present recent measurements of the triple differential cross sections for autoionizing states of helium in the *2l2l'* resonances.

Acknowledgments

This work was supported by the Scientific and Technological Research Council of Turkey (TUBITAK) through grant 109T738 and AKU-BAPK through grant 14.FENED.05.

References

[1] R. P. Madden and K. Codling 1965 *Astrophys. J.* **141**, 364.

[2] O. Sise, M. Dogan, I. Okur, A. Crowe 2010 *J. Phys. B: At. Mol. Op. Phys.* **43** 185201.

[3] O. Sise, M. Dogan, I. Okur, A. Crowe 2011 *Phys. Rev. A* **84 (2)** 022705.

[1] e-mail: mzekiler@aku.edu.tr

P26

Dalitz plot analysis of three-body fragmentation dynamics of CO_2^{q+} ($q = 3,\ 4$)

Enliang Wang[*,†], **Xu Shan**[*,†], **Zhenjie Shen**[*,†], **Maomao Gong**[*,†], **Yaguo Tang**[*,†], **Xiangjun Chen**[*,† 1]

[*] Hefei National Laboratory for Physical Sciences at Microscale and Department of Modern Physics, University of Science and Technology of China, Hefei, Anhui, 230026, China

[†] Synergetic Innovation Center of Quantum Information and Quantum Physics, University of Science and Technology of China, Hefei, Anhui 230026, China

Synopsis Electron impact induced three-body fragmentation dynamics of CO_2^{q+} ($q = 3,\ 4$) is investigated. The fragmentation dynamics of each dissociative channel is analyzed by Dalitz plot. For CO_2^{3+}, linear fragmentation, molecular bending and sequential fragmentation mechanisms are observed. It is found that two dissociative channels contribute to the fragmentation of CO_2^{4+} and the fragmentation dynamics of these two channels are quite different from each other.

The fragmentation dynamics of multiple ionized molecule has captured much interests recently [1, 2]. In this work, we present the Dalitz plot analysis [3] of the fragmentation dynamics of highly ionized molecule, CO_2^{q+} ($q = 3, 4$). The following three dissociative channels are analyzed:

$$
\begin{aligned}
CO_2^{3+} &\rightarrow C^+ + O^+ + O^+, &(1)\\
CO_2^{4+} &\rightarrow C^{2+} + O^+ + O^+, &(2)\\
&\rightarrow C^+ + O^+ + O^{2+}. &(3)
\end{aligned}
$$

In Dalitz plot, each point represent a specific momentum correlation among the three fragments. As shown in Figure 1 (a), the events located in the bottom center of the plot (in red dashed oval) originate from direct linear fragmentation. The events located on the y axis (marked by a black oval) originate from molecular bending dissociation. The events marked by "X" curve originate from sequential fragmentation.

The experimental Dalitz plots of channels (1), (2) and (3) are shown in Figure 1 (b), (c) and (d), respectively. For channel (1), all of linear fragmentation, molecular bending and sequential fragmentation mechanisms are observed. For channels (2), as shown in Figure 1 (c), most of the data locate symmetrically along y-axis and the most intense area locate near (0, -1/3) indicating that the linear fragmentation dominantly and a little bit dissociate through molecular bending fragmentation in this channel. Different from channel (2), as shown in Figure (d), most of the data locate at bottom right of the Dalitz plot for channel (3) indicating the asynchronous breakup mechanism predominantly in this channel. In this situation, the two bonds of CO_2^{4+} break simultaneously at which moment C^+-O^{2+} bond stretches much more than C^+-O^+ bond.

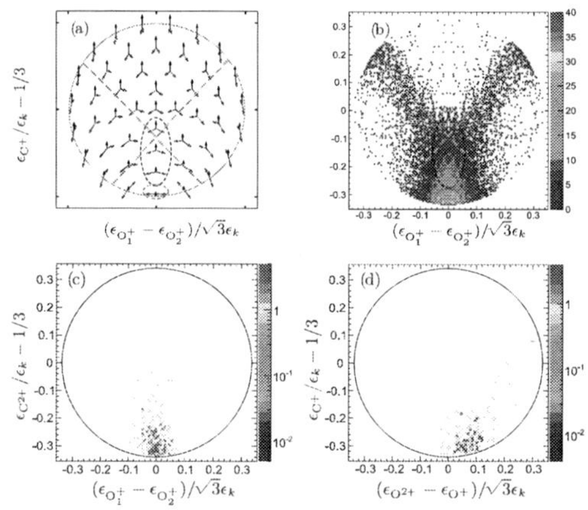

Figure 1. Dalitz plots for three-body fragmentation of CO_2^{q+} ($q = 3,\ 4$). (a) Momentum correlation features for various points in the Dalitz plot. (b) Channel (1). (c) Channel (2). (d) Channel (3).

References

[1] N. Neumann *et al* 2010 *Phys. Rev. Lett.* **104** 103201

[2] C. Wu *et al* 2013 *Phys. Rev. Lett.* **110** 103601

[3] R. H. Dalitz 1953 *Philo. Mag.* **44** 1068

[1]E-mail: xjun@ustc.edu.cn

P27

Three-body fragmentation dynamics of OCS^{4+} investigated by 500 eV electron collision

ZhenJie Shen*†, MaoMao Gong*†, EnLiang Wang*†, Xu Shan*†, XiangJun Chen*† 1

* Heifei National Laboratory for Physical Sciences at Microscale and Department of Modern Physics, University of Science and Technology of China, Hefei, Anhui, 230026, China

† Synergetic Innovation Center of Quantum Information and Quantum Physics, University of Science and Technology of China, Hefei, Anhui, 230026, China

Synopsis The three-body fragmentation dynamics of OCS^{4+} investigated by electron collision at impact energy of 500 eV is reported in this abstract. By analyzing the Dalitz plot, it is found that OCS^{4+} mostly dissociate through direct breakup together with a bit of sequential fragmentation where C-S bond dissociate firstly and then O-C bond dissociate after a time of delay. The fragmentation dynamics of various mechanisms are presented using Newton diagrams.

The non-sequential and sequential fragmentation dynamics of small molecules has captured interests in recent years [1, 2, 3]. For non-sequential fragmentation the chemical bonds break simultaneously. On the other hand, however, the molecular bond break one after another through sequential fragmentation. In this work, the electron impact fragmentation dynamics of OCS^{4+} is studied at impact energy of 500 eV,

$$e^- + \mathrm{OCS} \rightarrow \mathrm{OCS}^{4+} + 5e^-. \quad (1)$$

The main dissociative channel of OCS^{4+} is investigated,

$$\mathrm{OCS}^{4+} \rightarrow \mathrm{O}^+ + \mathrm{C}^+ + \mathrm{S}^{2+}, \quad (2)$$

where the three ions are detected in coincidence in experiment.

The Dalitz plot of channel (2) is shown in Figure 1 (a). The most intense areas locate near the bottom center of the plot which indicate that the direct breakup dominantly in the dissociation process of OCS^{4+}. In this situation O-C and C-S bond break simultaneously. Besides the direct breakup process, there are also some events locate along the diagonal of the Dalitz plot indicating that OCS^{4+} has some probability dissociate via sequential fragmentation. In this case, the C-S bond dissociate in first step,

$$\mathrm{OCS}^{4+} \rightarrow \mathrm{OC}^{2+} + \mathrm{S}^{2+} \rightarrow \mathrm{O}^+ + \mathrm{C}^+ + \mathrm{S}^{2+}. \quad (3)$$

After a time of delay, O-C bond dissociate in the second step.

The Newton diagrams of total, direct breakup and sequential fragmentation mechanisms are shown in Figure 1 (b), (c) and (d), respectively. The circular structures in Figure 1 (b) and (d) clearly represent the sequential mechanism.

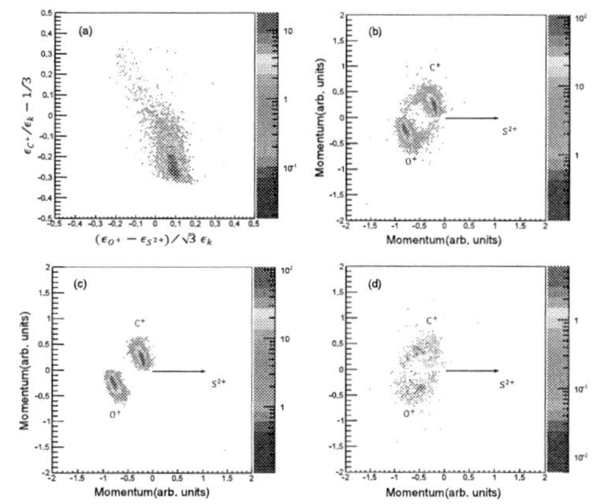

Figure 1. Dalitz plot of three-body fragmentation of OCS^{4+} (a) and Newton diagrams for total (b), direct breakup (c) and sequential fragmentation (d).

References

[1] N. Neumann *at al* 2010 *Phys. Rev. Lett.* **104** 103201

[2] C. Wu *at al* 2013 *Phys. Rev. Lett.* **110** 103601

[3] B. Wales *at al* 2014 *J. Electron Spectrosc.* **195** 332

[1]E-mail: xjun@ustc.edu.cn

P28

Correlation and quantum entanglement in Rydberg states of the helium atom

Yen-Chang Lin*,# 1, Te-Kuei Fang† 2, Yew Kam Ho# 3

*Graduate Institute of Applied Science and Engineering, Fu-Jen Catholic University, New Taipei City, Taiwan
† Department of Physics, Fu Jen Catholic University, New Taipei City, Taiwan
# Institute of Atomic and Molecular Sciences, Academia Sinica, Taipei, Taiwan

Synopsis Entanglement entropies for the singly excited $1snl\ ^{1,3}S^e$, $^{1,3}P^o$, $^{1,3}D^e$, and $^{1,3}F^o$ states in the helium atom are quantified by using configuration interaction with *B*-spline basis wave functions.

Entanglement is fundamental phenomenon in quantum physics, and it is related to correlation and interactions among subsystems within a whole system. Here, we use von Neumann entropy (see [1, 2] and references therein) and linear entropy [3] to quantify classical and quantum correlation. We calculate von Neumann entropy $S_{vN}(\rho_A) = -Tr(\rho_A \log_2 \rho_A)$ and linear entropy $S_L(\rho_A) = 1 - Tr(\rho_A^2)$ in two-electron atomic systems, with $\rho_{AB} = |\varphi_{AB}\rangle\langle\varphi_{AB}|$ being the density matrix. In atomic systems, particles such as electrons are identical particles, and one-particle reduced density matrix is defined as $\rho_A = \rho_B = Tr(\rho_{AB})$. The linear entropy S_L is an approximation of von Neumann entropy S_{vN}. An alternate definition [4] for quantum entanglement was given as $E(\rho) = |S(\rho) - S(\rho^{(0)})|$, where $\rho^{(0)}$ is for non-interacting and non-entangled states. It is a classical correlation limit. Usually, for ground states we have $S_{vN}(\rho^{(0)}) = 0$, $S_L(\rho^{(0)}) = 0$, and for the two electrons in an excited state, $S_{vN}(\rho^{(0)}) = 1$ and $S_L(\rho^{(0)}) = 0.5$. In our previous calculations, states with total angular momentum quantum L=0 [1, 3] in helium-like systems were investigated. In the present work, singly excited higher angular momentum ($L > 0$) states, such as the $1snl\ ^{1,3}P^o$, $^{1,3}D^e$ and $^{1,3}F^o$, as well as $^{1,3}S^e$ states, are calculated. The wave functions are obtained by using configuration interaction with *B*-spline basis with sizes ranging from 7000 to 9600. Our results indicate that states with different atomic structure will have different entanglement entropies. Figure 1 shows the energy difference (relative to the ground state of helium) *vs* effective quantum numbers n^* of the $^1P^o$ and $^3P^o$ states, in log-log plot. Figure 2 shows the counterpart for entanglement *vs* n^*, in log-log plot, indicating entanglement diminishes as the excited electron moving away from the nucleus. Details of our results will be presented at the meeting.

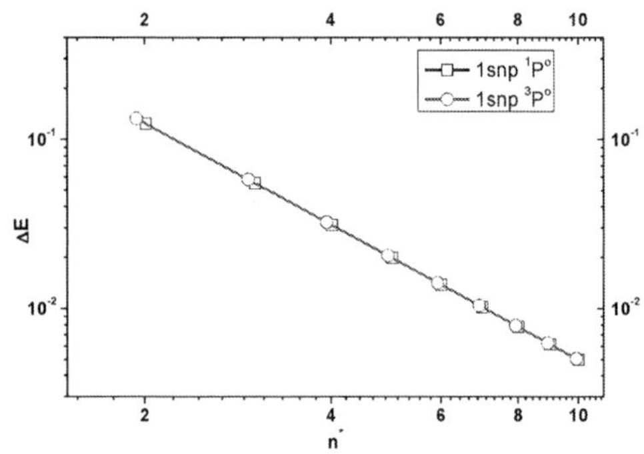

Figure 1. Energy difference relative to the ground state *vs* n^*, in log-log plot.

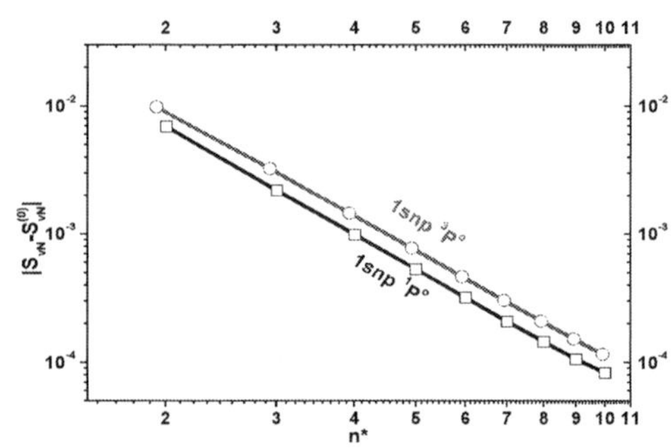

Figure 2. Entanglement von Neumann entropy *vs* n^*, in log-log plot.

Acknowledgement: This work was supported by MOST of Taiwan.

References

[1] Y.-C.Lin and Y. K. Ho 2015 *Can. J. Phys.* online published. doi: 10.1139/cjp-2014-0437
[2] Y.-C. Lin, T.K. Fang and Y. K. Ho 2015 *Phys. Plamas* **22** 032113
[3] Y.-C. Lin, Y.-C. Lin and Y. K. Ho 2013 *Phys. Rev. A* **87** 022316
[4] G. Benenti, S. Siccardi and G. Strini 2013 *Eur. Phys, J. D* **67** 83

[1] E-mail: linyc@pub.iams.sinica.edu.tw
[2] E-mail: 051420@mail.fju.edu.tw
[3] E-mail: ykho@pub.iams.sinica.edu.tw

P29

Effect of orthogonalization on total ionization cross sections by electron impact :Application to small molecules

S. Nehaoua[*], S. Houamer[*,1], C. Dal Cappello[$], M. Chinoune[*], A. Galstyan[§] and A. C. Roy[£]

[*] LPQSD, Department of physics, Faculty of Science, University Sétif-1, 19000, Setif, Algeria
[$] SRSMC, UMR CNRS 7565, Univ. of Lorraine, BP 70239, 54506 Vandoeuvre-les-Nancy, France
[§] Faculty of Physics, Lomonosov Moscow State University, Moscow 119991, Russia
[£] School of Mathematical Sciences, Ramakrishna Mission Vivekananda University, Belur Math 711202, West Bengal, India

Synopsis Total ionization cross sections by electron impact are calculated for some small molecules by using an improved first Born approximation which has been previously applied for atomic targets by Bartlett and Stelbovics [1]. A full orthogonalization of the final state to the initial state has been performed to evaluate the cross sections. The results obtained in the present model are immensely improved when compared with the first Born model without orthogonalization.

Differential and total cross sections for the ionization of atoms and molecules by electron impact provide fundamental data in a wide range of applications. Besides their strategic applications, investigations of molecular targets constitute effectively a challenging task for theorists because of their multicenter character. We present here an improved first order model (FBA-OCW) based upon the single particle picture and where the coulomb wave representing the final state is orthogonalized to all occupied molecular initial states of the target [2]. The molecular orbitals of the target are sometimedescribed by single center slater type wave functions .

In Fig. 1 we show the TCS for the ionization of H2O in an incident energy range from threshold to 5 keV . It is seen that calculations without orthogonalization (dashed line) largely overestimate the experiments at low impact energy especially in the peak region. When orthogonalization is used, the results are considerably improved. At low impact energy our theoretical results reproduce very well the data and the amplitude is quite well reproduced when compared with experiments. The peak is also well located. The inner orbital 2a1 is the most sensitive to the orthogonalization, the amplitude is practically divided by four while for the other orbitals the improvement over the traditional first Born model without orthogonalization (FBA-CW) which overestimates the TCS at low impact energies. In conclusion, the FBA-OCW is capable of reproducing the amplitude at low incident energy highlighting an overall good agreement with experiments with the peaks somewhat reasonably reproduced and well located

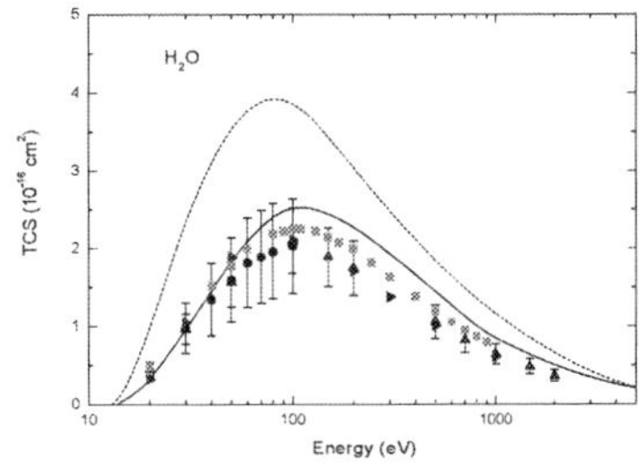

Figure 1. Comparison of the measured data with our theoretical total cross sections versus incident energy with the FBA-OCW (solid line) and FBA-CW (dashed line) in the case of H2O.

References

[1] P L Bartlett and A T Stelbovics, Phys. Rev. A **66**, 012707 (2002)

[2] S. Nehaoua et al ., Eur. Phys. J. D **69**, 86 (2015)

s_houamer@univ-setif.dz

P30

Non-perturbative B-spline R-matrix with pseudo-states calculations for electron-impact excitation-ionization of helium to the n=3 states

Oleg Zatsarinny [1] **and Klaus Bartschat** [2]

Department of Physics and Astronomy, Drake University, Des Moines, Iowa 50311, USA

Synopsis We present fully-differential cross sections for electron-impact ionization plus simultaneous excitation of helium obtained from a non-perturbative close-coupling formalism with our B-spline R-matrix approach. Using a large number of pseudo-states we obtain excellent agreement with directly measured cross-section ratios for ionization leaving the residual He^+ ion in either the 1s ground state, the n = 2 (2s + 2p) excited states, or the n = 3 (3s + 3p + 3d) excited states.

Electron-impact ionization with simultaneous excitation is a highly challenging problem, both experimentally and theoretically. Taking e – He as the proto-type collision system, the challenges are due to the fact that all three electrons (the projectile and both target electrons) undergo significant changes in their respective quantum states. Hence the problem cannot be simplified further by effectively treating one of the target electrons as a "spectator" that remains in the 1s orbital.

In recent years, we have further developed our fully non-perturbative B-spline R-matrix (close-coupling) with pseudo-states (BSRMPS) method and applied it to both the direct ionization [1] and ionization with excitation problems [2,3]. Despite the success achieved in employing a projection approach to obtain the cross sections from excitation amplitudes of the positive-energy discrete (i.e., finite-range) pseudo-states, some skepticism remains about the validity of the approach. The method is indeed not exact [4], but it provides a practical interpolation scheme (currently the only one known) that should systematically improve when the density of pseudo-states is increased.

To check this hypothesis, we further optimized our parallelized BSR codes. This allowed us to increase the number of pseudo-states to 1,254 and thereby also consider ionization with excitation to the n = 3 states of He^+. Figures 1 and 2 show a comparison of our predictions with directly measured cross-section ratios [5] for ionization without excitation and ionization with excitation to He^+(3s+3p+3d). The agreement with the experimental data is excellent for both symmetric (Fig. 1) and asymmetric (Fig. 2) sharing of the excess energy between the two outgoing electrons. Normalizing to the well-known results for ionization without excitation will then allow for the determination of absolute cross sections for this process [6].

This work was supported by the United States National Science Foundation under grants PHY-1212450 and PHY-1430245, and the XSEDE allocation PHY-090031.

[1] E-mail: oleg.zatsarinny@drake.edu
[2] Email: klaus.bartschat@drake.edu

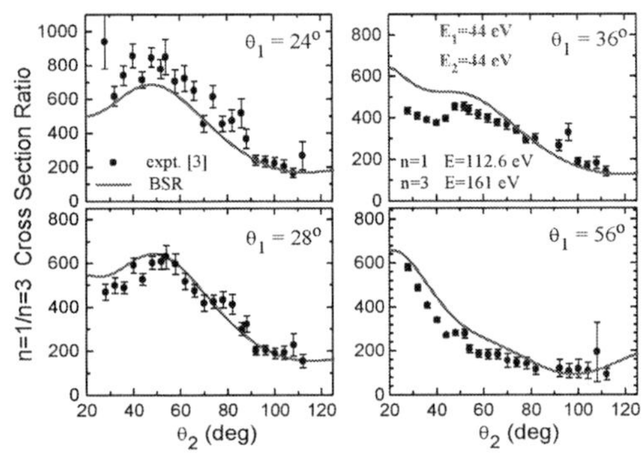

Figure 1. TDCS ratio for electron impact ionization of helium with the residual ion left in either the 1s ground state or the excited n = 3 states. The excess energy is shared equally between the two outgoing electrons.

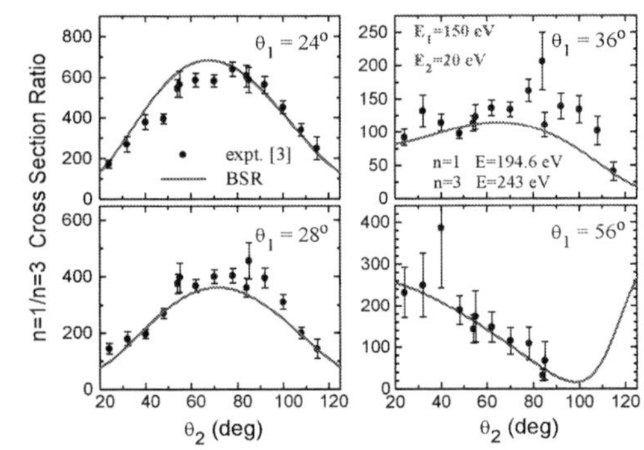

Figure 2. Similar to Fig. 1 for asymmetric sharing of the excess energy.

References

[1] O. Zatsarinny and K. Bartschat 2012 *Phys. Rev. A* **85** 062709
[2] O. Zatsarinny and K. Bartschat 2011 *Phys. Rev. Lett.* **107** 023203
[3] O. Zatsarinny and K. Bartschat 2014 *J. Phys. B* **47** 061001
[4] I. Bray *et al* 2014 *Phys. Rev. A* **90** 022710
[5] S. Bellm *et al* 2007 *Phys. Rev. A* **75** 042704
[6] S. Bellm *et al* 2008 *Phys. Rev. A* **78** 032710

P31

Mechanisms of Chiral Sensitivity in Electron-Molecule Interactions

J.M. Dreiling* [1], **F.W. Lewis**†, **and T.J. Gay***

*Department of Physics and Astronomy, University of Nebraska-Lincoln, Lincoln, Nebraska 68588-0299, USA
† Department of Chemical and Forensic Sciences, Northumbria University, Newcastle upon Tyne, NE1 8ST, UK

Synopsis We have studied the interactions between longitudinally-polarized (chiral) electrons and various gas-phase chiral molecules. Three molecules and two interactions, dissociative electron attachment and electron transmission, were considered. The rates of both interactions depend on the sign of the incident electron helicity (chirality) for a given target handedness.

We present the results of our search for asymmetric interactions between longitudinally spin-polarized electrons and different chiral halocamphor molecules. The asymmetry is defined as

$$A = \left(\frac{I\uparrow - I\downarrow}{I\uparrow + I\downarrow}\right)_L - \left(\frac{I\uparrow - I\downarrow}{I\uparrow + I\downarrow}\right)_R,$$

where $I_\uparrow$ ($I_\downarrow$) is the current measured for spin-up (spin-down) electrons and the "L" and "R" subscripts correspond to the left- and right-handed chirality of the molecules [1]. Two electron-molecule interaction channels were studied: electron transmission (related to the total scattering cross section) and dissociative electron attachment (DEA). Three halocamphor molecules were investigated: 3-bromocamphor ($C_{10}H_{15}BrO$), 3-iodocamphor ($C_{10}H_{15}IO$), and 10-iodocamphor.

While the transmission asymmetry data do not show a strong molecular dependence, the DEA asymmetries collected for bromocamphor and iodocamphor are qualitatively different, suggesting that the atomic number of the heaviest atom in the molecule plays a crucial role in the asymmetric interactions. The DEA asymmetry data for 3- and 10-iodocamphor have the same qualitative behavior, but the 10-iodocamphor asymmetry is about twice as large at the lowest energies investigated, so the location of the heavy atom in the camphor molecule also affects the asymmetries.

References

[1] J.M. Dreiling and T.J. Gay 2014 *Phys. Rev. Lett.* **113** 118103.

[1] E-mail: jmdreiling2@gmail.com

P32

Kinematically complete low-energy (e, 2e) study of neon: Internormalized triple-differentical cross sections in 3D kinematics

XueGuang Ren*‡ [1], Sadek Amami§, Oleg Zatsarinny†, Thomas Pflüger*‡, Marvin Weyland*‡, Woon Yong Baek*, Hans Rabus*, Klaus Bartschat†, Don Madison§, Alexander Dorn‡

* Physikalisch-Technische Bundesanstalt, 38116 Braunschweig, Germany
‡ Max-Planck-Institute for Nuclear Physics, 69117 Heidelberg, Germany
§ Physics Department, Missouri University of Science and Technology, Rolla, Missouri 65409, USA
† Department of Physics and Astronomy, Drake University, Des Moines, Iowa 50311, USA

Synopsis Low-energy ($E_0 = 65$ eV) electron-impact ionization of Ne (2p) is studied to thoroughly test state-of-the-art theoretical approaches. The experimental data are compared to predictions from a hybrid second-order distorted-wave Born plus R-matrix approach (DWB2-RM), the distorted-wave Born approximation with inclusion of post-collision interaction (DWBA-PCI), a three-body distorted-wave approach (3DW), and a B-spline R-matrix (BSR) with pseudostates approach. Excellent agreement is found between experiment and the 3DW and BSR theories. The importance of PCI effects is clearly visible in this low-energy electron-impact ionization process.

The dynamics of electron-impact ionization of simple systems such as atomic hydrogen and helium have now been well understood [1, 2, 3]. Recent studies for the ionization of neon by 100 eV electron-impact showed an unprecedented agreement between experiment and BSR predictions [4]. The physical effects of PCI as well as electron exchange and charge-cloud polarization in the projectile-target interaction are expected to become even more pronounced with decreasing projectile energy. Here, we present a combined experimental and theoretical study on the low-energy ($E_0 = 65$ eV) electron-impact ionization of neon.

The experimental data were measured using a reaction microscope [5, 6], which can cover nearly the entire 4π solid angle for the secondary electron emission. The measured cross sections are internormalized across all different scattering angles (θ_1) and ejected energies (E_2), which provide a thorough test ground for theory. As one example, the (e, 2e) triple-differential cross sections (TDCS) for $\theta_1 = -12.5°$ and $E_2 = 2$ eV are presented in Figure 1 for the scattering plane and the full-perpendicular plane. The experimental data are compared to various theoretical predictions from the 3DW, BSR, DWB2-RM, DWBA-PCI models. Excellent agreement is found between experiment and the 3DW and BSR theories. Significant discrepancies between DWB2-RM and experiment are observed near the projectile forward direction, while the DWBA-PCI model provides a clear improvement over the DWB2-RM calculations in this angular range. This indicates that PCI effects play a very important role in the low-energy ionization process studied here [7]. More results, including three-dimensional (3D) presentations of the TDCS, will be shown at the conference.

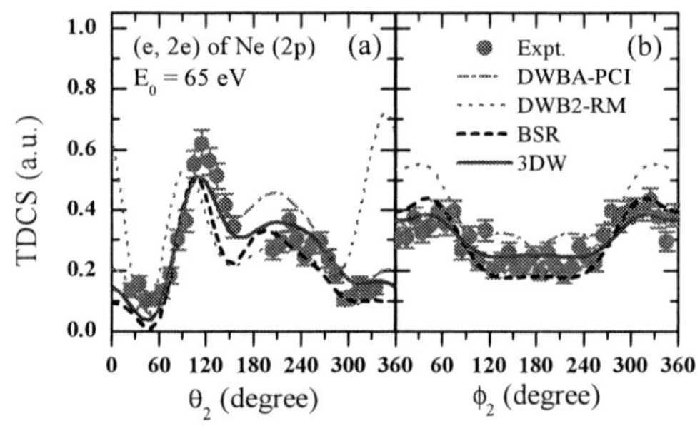

Figure 1. TDCS for the ionization of Ne ($2p$) presented as a function of the secondary electron (e_2) emission angle at $\theta_1 = -12.5°$ and $E_2 = 2$ eV. (a): TDCS in the scattering plane; (b): TDCS in the plane perpendicular to the incident beam direction.

References

[1] T. N. Resigno *et al* 1999 *Science* **286** 2474
[2] X. Ren *et al* 2011 *Phys. Rev. A* **83** 052711
[3] O. Zatsarinny *et al* 2011 *Phys. Rev. Lett.* **107** 023203
[4] T. Pflüger *et al* 2013 *Phys. Rev. Lett.* **110** 153202
[5] J. Ullrich *et al* 2003 *Rep. Prog. Phys.* **66** 1463
[6] X. Ren *et al* 2014 *J. Chem. Phys.* **141** 134314
[7] X. Ren *et al* 2015 *Phys. Rev. A* **91** 032707

[1]E-mail: xue.g.ren@ptb.de

High resolution electron ejected spectra of He, Ne and Ar by high energy electrons

B. P. Marinković [1], **J. J. Jureta, and A. R. Milosavljević**

Institute of Physics, University of Belgrade, Pregrevica 118, 11080, Belgrade, Serbia

Synopsis A systematic study of autoionisation region of rare gas atoms (He, Ar and Ne) by high energy electrons was carried out at ejection angles from 10° to 130°. The best obtained energy resolution of the ejected electron spectra was 45 meV. The high resolution and high sensitivity allowed us to see separated singlet from triplet states at high incident energies. Comparison with previous works gives a good agreement in energy positions of measured features.

The autoionising region of rare gas atoms has been studied extensively in the past by electron, photon and ion collisions, nevertheless the wide scale ejected electron spectra with high resolution that include angular analysis have not been systematically presented. We started with the measurements of He [1], Ne [2] and Ar [3,4] atoms at the high excitation electron energies from 500 eV to 2000 eV and span the ejected electron energies in the broad range (see Table 1), so that we were able to cover the regions of single and double excited autoionising states.

The experimental set-up consists of *a*) High energy electron gun mounted at the turn table that enables rotation from 10° to 130° around analyzer axis *b*) Gold plated platinum–iridium tube 30 mm long with the inside diameter of 0.5 mm used as a source of effusive beam of rare gases; *c*) High resolution electron analyzer with 7 channeltrons as detectors. The obtained energy resolution of ejected spectra was from 45 to 80 meV depending of the analyzer tuning. This is enough to resolve the singlet-triplet splitting in the ion core in Ne (0.098 eV) and Ar (0.177 eV).

Table 1. Ionisation energies and the range of measured ejected electron energies in eV.

	Ionisation (eV)	Ejected electron energies (eV)
He	24.59	32.0-55.0
Ne	21.56	12.0-46.0
Ar	15.76	5.5-22.5

The spectra are shown with subtracted background and without normalisation of the obtained data. The calibration point for ejected electron energy scale was taken from well known helium $(2s2p)^1P$ state in autoionisation region at 35.54 eV (60.13 eV) [5].

A series of ejected electron spectra was obtained for all three target atoms performing the systematic variation of incident energies and ejected angles. The spectra are rich in features, many of them not identified earlier. Further theoretical investigation is needed in order to assign the configuration of observed states.

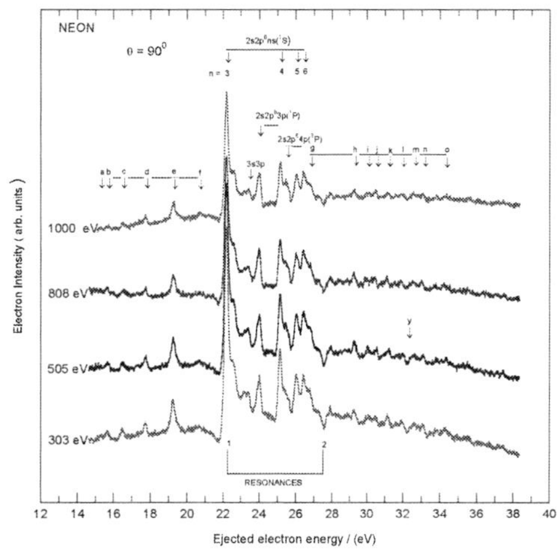

Figure 1. Ejected electron spectra of neon obtained at an ejection angle of 90° and incident electron energies or excited energies of 303, 505, 808 and 1000 eV respectively shown on the left hand side on the figure. The energy region of ejected electrons is from 14 to 38 eV or 35.56 to 59.56 eV of excitation energy with energy width per channel of 0.020 eV. Small arrows on the top of the figure show energy positions of the features, while vertical lines at the bottom of the figure mark positions of resonances.

References

[1] J. J. Jureta *et al* 2014 *Int. J. Mass. Spectrom.* **365-366** 114
[2] J. J. Jureta *et al* 2015 *Eur. Phys. J. D* **69** 74
[3] J. J. Jureta *et al* 2014 *Proc. Int. Conf. SPIG* **27** 50
[4] B.P. Marinković *et al* 2015 *ICPEAC* **29** 231
[5] R. P. Madden and K. Codling, 1963 *Phys. Rev. Lett.* **10** 516

[1] E-mail: bratislav.marinkovic@ipb.ac.rs

TIME SCHEDULE

July 29th		July 30th		July 31st		August 1st	
	8.00h - 8.45h	Registration					
	8.45h - 9.00h	Opening					
	Chairmen	*Díez Muiño*		*Kabachnik*		*Madison*	
	9.00h - 9.45h	Cederbaum		Kheifets		Stefani	
	Chairmen	*Bartschat*	*Piancastelli*	*Dorn*	*Bray*	*Schulz*	*Bolognesi*
	10.00h - 10.30h	D. Fischer	Sokell	Schulz	Rudenko	Madison	Avaldi
	10.30h - 11.00h	Hargreaves	Ueda	Zhang	Nakajima	Chen	Nahon
	11.00h - 11.30h	Coffee Break					
	Chairmen	*Colgan*	*Sokell*	*Takahashi*	*Mazza*	*Ancarani*	*Avaldi*
	11.30h - 12.00h	Bartschat	Furst	Yamazaki	Smirnova	Dorn	Nicolas
	12.00h - 12.20h	Ren	Amusia	Piwiński	Jiménez-Galán	Gervasoni	Pavlyukh
	12.20h - 12.40h	Wei	Scrinzi	Dong	Liu	Tõkési	Iablonskyi
	12.40h - 13.00h	Tashenov	Popov		Zhu	Farewell	Farewell
	13.00h - 15.15h	Lunch Break					
	Chairmen	*Chen*	*Ueda*	*Landers*	*Grum-Grzhimailo*		
	15.15h - 15.45h	Paripas	Palacios	Stauffer	Mazza		
	15.45h - 16.15h	Martin	Schnorr	Surzhikov	Kazansky		
	16.15h - 16.35h	Koval	Travnikova	Nagy	Argenti		
	16.35h - 16.55h	Morini	Ayuso	Ho	Püttner		
		Coffee Break					
	Chairmen	*Paripas*	*Palacios*	*Stauffer*	*Scrinzi*		
Registration	17.30h - 18.00h	Colgan	Guillemin	Landers	Vasyutinskii		
OPEN at Miramar Palace	18.00h - 18.30h	Ancarani	Fang	Fojón	A. Fischer		
from 17.30h to 19.30h	18.30h - 20.00h	POSTER SESSION & DRINKS					
	20.30h --			CONFERENCE DINNER			
		↑ Electrons/Ions	↑ Photons	↑ Electrons/Ions	↑ Photons	↑ Electrons/Ions	↑ Photons